Interplanetary Travel

Travel

– An astronomer's guide

Dr. Sten Odenwald

The Astronomy Cafe

Praise for Sten Odenwald

"With his first book, *The Astronomy Café*, Sten Odenwald demonstrated that he belongs at the interface between the cosmic frontier and the public inquiry of that frontier."– Neil De Grasse Tyson (American Museum of Natural History).

"Odenwald not only sets a comfortable conversational tone; he adds a sense of humanity that some science books tend to omit" - Sky and Telescope

"Odenwald's concise writing and his sense of humor and layperson's writing style is a laudable public service" – Bloomsbury Review

Other books by this author:

Interplanetary Travel: Prospects and challenges
Interstellar Travel: An Astronomer's Guide
Eternity: A User's Guide
Exploring Quantum Space
Ask the Astronomer
Solar Storms

The 23rd Cycle
Patterns in the Void
The Astronomy Café
Back to the Astronomy Café
Stepping Through the Stargate

For related information about astronomy
visit the Astronomy Café website at
http://www.astronomycafe.net

Copyright © 2015, Sten Odenwald

Acknowledgments

I would very much like to thank the many readers of my Huffington Post Blogs who made so many excellent comments about my 'Interplanetary Travel' essays, and who inspired me to write this more-comprehensive guide!

I would especially like to thank Robert Heinlein, Isaac Asimov, Arthur C. Clarke, James Michner, Greg Bear, Stephen Baxter and Gregory Benford, who inspired my dreams, even as I learned to deal with the technical and scientific realities of space travel.

Cover art: A concept image of a spacecraft powered by a fusion-driven rocket. In this image, the crew would be in the forward-most chamber. Solar panels on the sides would collect energy to initiate the process that creates fusion. Travel times to distant planets like Neptune would take a few weeks. Credit: Prof. John Slough and the Fusion Driven Rocket team at the University of Washington. The work was funded under a NASA Innovative Advanced Concepts grant. A video giving more details about the design and operation can be found here:

https://www.youtube.com/watch?v=tHSOmOu61b0

Table of Contents

Prolog

Interplanetary travel?

No one ever thought it would be this hard, especially the folks that are not literally 'rocket scientists'! Why is it that 50 years after the official start of the Space Age we do not have colonies on the moon and Mars already? Why are we still plodding along in Earth orbit? There are a million of these kinds of questions you can ask, but when you dig into their answers you uncover something very odd indeed.

In my previous book '*Interstellar Travel: An Astronomer's Guide*' I laid out all of the astronomical challenges to interstellar travel and colonization. The bottom line was that this was an adventure filled with dreams but no technology to get us there. In this book, '*Interplanetary Travel: Prospects and challenges from an astronomer's perspective*', we discover that in fact we have all the technology we need to make it happen, but our current problem is that we lack dreams. Think about it.

Most of the science fiction you have read is all about interstellar travel. Our little neighborhood in the universe is almost always given short shrift. We have no dreams about our own solar system that can stand toe-to-toe with the technological fantasy of interstellar empires and exploration. We have no dreams about planetary colonization. There are no stories to entice us to invest in this effort; no great expectations about what we will find that is worth going after.

So as an astronomer, I look at interplanetary travel as very much the odd duck. Time and again NASA has developed technologies to make this exploration possible. And time and again, Congress has forbidden NASA from pulling out the stops to deploy the technology already developed to make it happen in our lifetimes. Nuclear rocket engines to take us to Mars in 30 days? We have those, but Congress forbids launching nuclear reactors into space because it is seemingly too expensive or potentially hazardous. Interplanetary laser communication and a gigabyte internet? We have that too, but Congress only lets us use it inside geosynchronous orbit!

These are just a few of the frustrations about making interplanetary travel a reality. It's not that we don't have the technology. The problem is Congress can't see the point in the investment and rushed timetable. Yet engineers keep plugging ahead, literally on their lunch-hours and coffee breaks, and developing what is needed on their benches and in their laboratories. Funded by their own dreams, and small pots of money cobbled together from government and private grants, they steadily advance the limits of what we can do.

In this book, I want to show you some of the best ideas we have about interplanetary travel. Does it really make sense to travel outside the asteroid belt, when all you will find there are moons made of ice? There is an entire mythology that has grown up in science fiction that has human miners spread across the solar system, but the astronomical reality is that beyond Mars, all you will find is ice. Mining the ice on the moons of Saturn is far more expensive than on the moons of Jupiter, so astronomy limits our economic activities to basically the inner solar system!

This book is not just about limits, but about what kinds of considerations go into making interplanetary travel an economic reality. In the end, you will discover that our future in space will depend on some very hard thinking about how we see ourselves interacting with the universe in the near future. History has shown us again and again that there is never a 'better time' to embark on journeys of exploration. The current investment in space exploration is only 0.6% of the annual federal budget in the US, not the 25% that some misinformed citizens seem to believe. We will not end poverty before we establish our first colony on Mars, but we may well be impoverished as a civilization if we decide such a challenge should not be met! Nevertheless, we still have to provide a reason for going beyond Earth orbit that makes sense.

Most of the reasons we can think of are pretty sound, but do not provide a compelling need that has to be met in the 21st century. There is no resource we can harvest in space that is more cost-effective to recover than from Earth's surface, or even on the ocean floor. Calls for exploration based on scientific curiosity are naive, given that the vast majority of the human population does not see the advancement of scientific knowledge as a significant, personal issue. Finding a politically or commercially-compelling reason for going into deep space is the single most difficult issue that is holding back interplanetary travel at a technological scale that would make it both cheap, safe, and fast. Even President Kennedy had no real use for space exploration because he considered it very expensive and not something the general voting public would ever support. Nonetheless, with the Soviet Union launching satellites, he realized that for purely political reasons the United States could not be perceived as a second-class power against the massive communist advancements around the world. Unless we find an equally compelling reason for interplanetary travel, Congress will never allocate more than NASA's current budget for such an undertaking.

This book will cover many of the aspects of interplanetary travel, such as why we should attempt it, where we will go, how much it will cost, the political and commercial climate, and of course the state of our technology today to undertake such a massive program.

Science Fiction

One of my earliest introductions to how interplanetary travel would work comes from Robert Heinlein's wonderful books 'Red Planet' (1949), 'Between Planets' (1951), 'The Rolling Stones' (1952) and 'Have Spacesuit. Will Travel' (1958). Heinlein was a former engineer well-versed in matters of rocketry. His universe consisted of very large colonies on Mars and Venus, and settlements on Ceres, Titan and the Galilean moons of Jupiter especially Ganymede. These were traditional domed communities that had matured into major settlements of upwards of 100,000 or more people. 'Red Planet' is set some time in the 21st century, with a major equatorial settlement and outlier colonies, and the beginnings of the terraforming process to make the Martian atmosphere breathable. Other stories like 'Podkayne of Mars' (1962) take place in the equivalent of 150,000-ton ocean liners (e.g. Queen Elisabeth II) in space that never touch down on a planet, but where people are shuttled back and forth on self-propelled, aerodynamic ships. These leviathans maintain a constant acceleration of 0.1Gs, so interplanetary travel takes about a few weeks to get from Mars to Earth, and the ships rotate to produce Earth-like artificial gravity. In 'Have Spacesuit:Will Travel' we hear that a ship can reach Pluto in three weeks at a 1-G acceleration for half the trip and 1-G deceleration for the final half. In Heinlein's universe, typical tourist travel between planets use what appear to be conventional 'reaction mass' rockets in 'The Rolling Stones', though the designs are similar to V-2 rockets, and so there is some considerable 'magic' involved in the engine design.

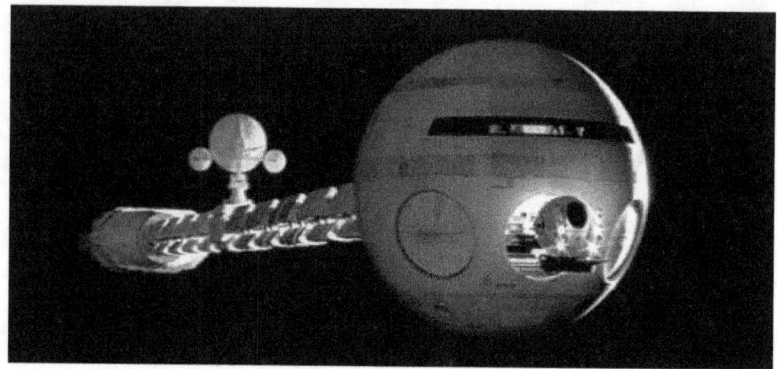

Arthur C. Clarke's *'2061:Odyssey three'* (1987) shows us a solar system where travel times are also in weeks not years, with a colony on Mars and major lunar settlements numbering close to one million people. Passenger travel to Saturn is a reality, with ships under constant 0.1-G acceleration touring the solar system.

By the 23rd Century, the *Babylon 5* universe (1995) created by J. Michael Straczynsky has the Mars colonies seeking independence from Earth, two large lunar colonies, a station orbiting the Jovian moon Io (*'Chrysalis'*), as well as colonies on Ganymede (*'Messages from Earth'*) and Europa (*'TKO'*). Travel from Io to Earth is several days by conventional ship with artificial gravity provided by rotation. Propulsion seems to be by ion engines powered by fusion reactors (e.g. Earthforce battle cruisers).

In the Star Trek (1966) universe of the 24th Century, we hear about the Martian star ship design lab at the colony on Utopia Planitia (1987: Star Trek:Next Generation), and the population of the lunar colonies is about 50 million (Star Trek:TNG:*A Time to Heal*). The Mars colonies seceded from Earth in the 22nd Century. Presumably travel times by 'impulse drive' (25% speed of light) are of the order of hours or days across the solar system using, as the story goes, deuterium (or the fantasy element trilithium!) fusion reactors that eject plasma as reaction mass, though we never see the incandescent material ejected by the engines.

Roger MacBride Allen's *The Ring of Charon* (1990) is set in a solar system centuries in the future where there are colonies on the Moon, Mars, Venus, Mercury, Titan and Ganymede, along with a large gravity research station orbiting Pluto's moon Charon. It is 16 days to Earth from this outpost. Then we have Stephen Baxter's universe of *'Manifold Space'* (2001) where we have a lunar colony but no outposts on Mars, however the technology needed for a 500-day trip to 1000-AU can be put into place very quickly once Humanity is motivated by alien invasion.

X-20 Dynasoar on top of an impressive rocket ca 1963. (Credit: Aerojet)

The more you read science fiction, the more you realize we are doing something very wrong. Interplanetary travel is inherently inexpensive and is something that a mid-sized company or an ambitious explorer can afford. In the Heinlein Universe, it is relatively easy to get a decent rocket ship that even allows you to be self-sustaining for a year or more. Prices seem to be in the millions of dollars for a good ship that can get you to the asteroid belt.

Miners become indentured to contracts to purchase these rockets, just as we have home mortgages to banks for our houses. If you want to step up to more luxury, in '*Enigma*' (1986) written by Michael Kube-McDowell, you can purchase a 'spaceliner' capable of round trips to the Jovian moons for about a couple of billion dollars. These ships can make the trip to Jupiter in one month each way, and keep its passengers happy and well-fed!

For rockets, which are the most common design in science fiction, conservation of momentum demands that if you want to go forward fast, you have to throw something away in the opposite direction very fast. This is called reaction mass, and the whole point of engine technology development is to make the mass ejection speed as close to the speed of light as you can, so that you can carry less and less fuel (making the payload-to-rocket mass ratio as high as possible). Interplanetary travel in science fiction has mastered this using 'nuclear ion propulsion'. These engines are so efficient that the propulsion mass is a small fraction of the mass of the rocket...so it is hardly ever mentioned. Look carefully at the design of the Babylon 5 spacecraft. Where are the hundreds of tons of reaction mass 'fuel' stored? How about the shuttle craft used in *Star Trek*? You don't store tons of reaction mass in those spiffy nacelles! And don't even get me started on Han Solo's Millennium Falcon, or Darth Vader's compact TIE-fighters! (TIE, by the way, is short for 'Twin Ion Engine'). Where's the beef? Very few science fiction stories talk about refueling their spacecraft, but a few cover designs can be found in the science fiction universe that are pretty interesting.

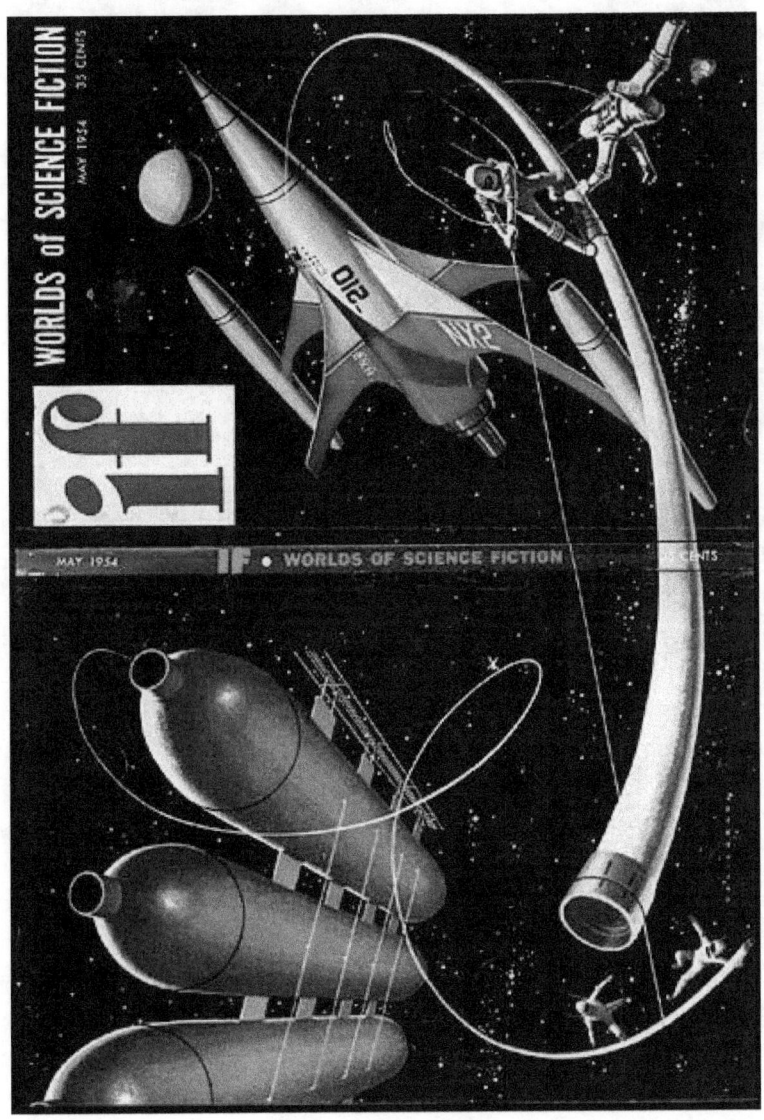

Even NASA is involved in this concept of refueling spacecraft in Earth orbit. Here's a version of a fuel depot designed by Boeing for near-Earth orbit. On-orbit fuel depots were included in NASA's initial Human Exploration Framework in 2010, which members of Congress approved. So, even for primitive chemical rockets, we can extend their reach into the solar system by setting up strategically-placed fuel dumps.

For the most part, the biggest hurdle to interplanetary travel is getting from place to place as fast as possible. We know that radiation effects and human physiology problems are made worse by prolonged trips, so this drives us to find faster and more efficient rocket technology. The good news is that we have many prototypes for these systems and several actually in operation. The bad news is that for lack of political will and dreams, we are seemingly hamstrung to actually get on with the actual use of these technologies in space.

Where should we go?

In the past, explorers had some general idea where they wanted to end up, and why they were willing to risk many hardships to get there. Christopher Columbus wanted to find a faster way to India to provide access to spices, without the arduous and costly journey overland or by ship around the treacherous Horn of Africa.

Simply put, very few major expeditions were launched purely for scientific curiosity. It was almost always the case that some economic, political or religious need was at the core of the action. Circumstances are no different today. The creation of the U.S civilian space program, NASA, and its goal of a manned landing on the moon by the end of the 1960's was almost completely driven by the Cold War need to demonstrate U.S technological, and therefore ideological, supremacy over the Soviet Union. Whatever science was accomplished was relatively secondary. To look beyond the moon and into the depths of the solar system, we have to completely re-think how we will approach the process of exploration, and what we expect to get in return.

Today we are 50+ years beyond the first human orbital adventures, and 40 years beyond the last moon landing, and have accumulated considerable sophistication in our basic scientific understanding

11

our solar system as exotic real estate. From the current catalog of planets, moons, asteroids and other objects with solid surfaces that are larger the 10 kilometers in diameter, (Jupiter, Saturn, Uranus and Neptune excluded), the total surface area of these objects is about 1900 million square kilometers. That works out to nearly four times the surface area (including oceans) of Earth. The planets Mercury, Venus, Earth and Mars and our Moon account for only 1226 million square kilometers or 65% of the total. The rest is found on the moons of Jupiter, Saturn, Uranus and Neptune (about 20%) in the Kuiper belt (about 14%) and the asteroid belt (about 1%). Note, although the asteroid belt has thousands of known bodies, only three are more than 200 km in radius, while 54 Kuiper Belt Objects are at least that big!

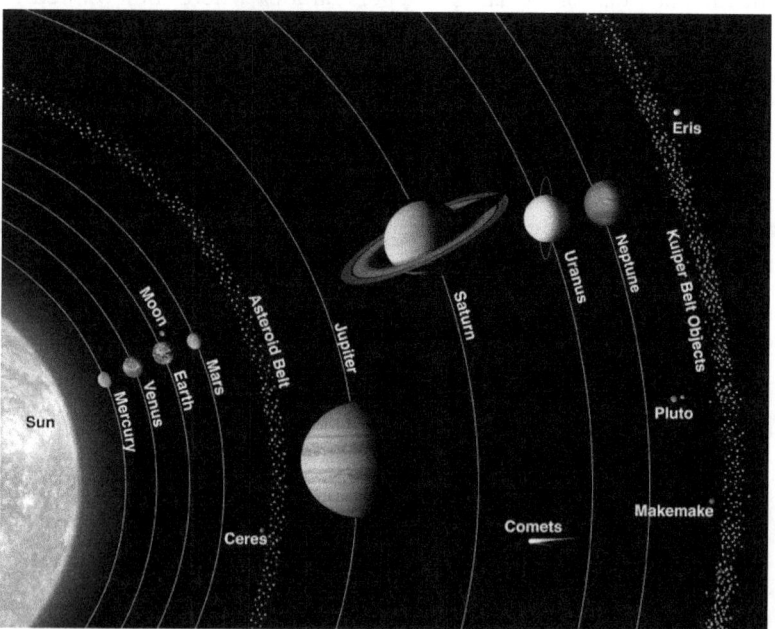

Although we can wrap our minds around the scale of the surface area of the solid bodies in our solar system, it is almost impossible to play a similar trick to understand its physical dimensions, from the sun to its most distant limits. There is no possible way to represent the vastness of our solar system in a single figure and get the scales of the planets and their orbits to look correct. This image is just a reference point to visually define the relative sizes of the planets, but not also their separations. (Image credit: NASA Space Place)

One practical way to think about the size of the solar system is to work with the time it takes a radio signal (or light) to get from one place to the other. If you have a colony living on a moon of Neptune, this will tell you what kind of one-way time delay you should expect between messages. To make this work, we will put all the planets in a row like the figure we have here, and create a table that includes all the possibilities for their minimum distances.

In statistics, if we have N objects selected M at a time, the number of unique combinations is just

$$C = \frac{N!}{M! \, (N-M)!} = \frac{9!}{2! \, 7!}$$

$$C = \frac{9 \times 8 \times 7 \times 6 \times 5 \times 4 \times 3 \times 2 \times 1}{7 \times 6 \times 5 \times 4 \times 3 \times 2 \times 1 \times 2 \times 1}$$

$$C = 36$$

For any two objects drawn from a collection of 9, there are 36 unique distances. The darkened boxes are redundant answers since the time from Earth to Venus is the same as the time from Venus to Earth and so on. The one-way times are in given hours. The distance between the sun and Pluto is 40 times the distance from Earth to the sun; a handy distance measure in the solar system called the Astronomical Unit (AU), which makes the Pluto distance 40 AU. The Earth, by definition, is in a near-circular orbit with an average distance of exactly 1 AU from the sun (149 million kilometers or 93 million miles). Since the speed of light is 300,000 km/sec, the Earth is 497 seconds or 8.3 minutes from the sun.

	Mer	Ven	Ear	Mar	Jup	Sat	Urn	Nep	Plu
Mer		0.05	0.09	0.16	0.67	1.3	2.6	4.1	5.5
Ven			0.04	0.11	0.62	1.2	2.6	4.1	5.4
Ear				0.07	0.58	1.2	2.5	4.0	5.4
Mar					0.51	1.1	2.5	4.0	5.3
Jup						0.6	2.0	3.4	4.8
Sat							1.4	2.9	4.2
Urn								1.5	2.9
Nep									1.4
Plu									

An interesting feature of our solar system is that the rocky 'terrestrial' planets Mercury through Mars are crowded into a zone only 170 million kilometers wide, however the spacing between Jupiter, Saturn, Uranus and Neptune varies from 650 million kilometers between Jupiter and Saturn, to 1.6 billion kilometers between Uranus and Neptune. The outer planets are far more massive than the terrestrial planets so they swept out much larger volumes of space along their orbits in order to grow so large.

Because of the much greater spacings in the outer solar system, the effort to travel between planets grows enormously more difficult among the outer planets. Also, time delays for communication increase from an annoying few dozen minute among the inner planets to many hours among the outer planets. While robotic probes tele-operated from Earth are still an option when exploring the inner planets, this mode is largely out of the question in the outer solar system where spacecraft must be far more autonomous.

Prior to the 1960s, our most powerful ground-based telescopes could only see a few of the solid-surface objects with any degree of clarity. Even so, the only object studied with adequate resolution to fully understand the character of its surface was our own moon. Telescopic studies at Yerkes Observatory (40-inch) and Mount

Wilson Observatory (10-inch) and assembled by Gerard Kuiper in his *Photographic Lunar Atlas* in 1960 had succeeded in resolving features as small as one kilometer on the half of the surface perpetually facing Earth.

Angular to linear size;

$$Angle = 206265 \frac{Diameter(km)}{Distance(km)}$$

Example for Mercury:
Diameter=3032 km
Distance= 77 million km
Angle= 8 arcseconds.

The challenge of studying solar system objects from Earth can be summarized in the following table, which indicates the angular diameter of the planet as viewed from Earth at the minimum-possible distance.

Object	Distance (km)	Diameter (arcsec)
Mercury	77 million	8
Venus	42 million	59
Mars	55 million	25
Ceres	240 million	0.8
Ganymede	588 million	1.8
Titan	1.2 billion	0.9
Titania	1.7 billion	0.2
Triton	2.7 billion	0.2
Pluto	4.3 billion	0.07

It is pretty obvious that Venus with its dense clouds is not a good observing target, but the miniscule angular sizes of the other objects makes them very difficult to study as well. Here is a recent telescopic image of Mercury. Using the 4.1-meter SOAR Telescope, astronomer Gerald Cecil

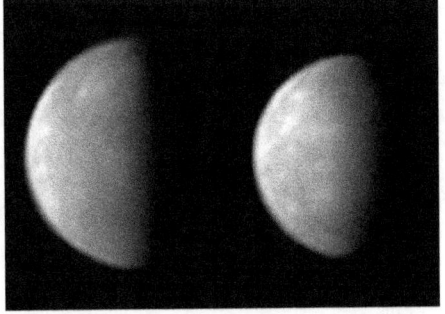

and undergraduate Dmitry Rashkeev managed to resolve details as small as 15 km across on March 23 and April 1, 2007. (Image credit: University of North Carolina/ G. Cecil)

Another popular telescopic target is, of course, Mars. The French observatory in the Pyrenees mountains, Pic du Midi, was specifically built to take advantage of the excellent seeing conditions to photograph and study planetary bodies. Here is a photograph of Mars taken during its 1992 opposition (closest point) to Earth. The resolution in this image is about 20 kilometers. (Image credit: Pic-du-Midi Observatory)

As for other objects in the solar system, the imaging from Earth is even worse. The moon of Jupiter, Ganymede, seen in this image, has an angular size of just under 2 arcseconds or 1/12 the diameter of Mars. This image was taken by photographer Damian Peach using a ground-based 30-cm telescope and sophisticated image processing. The angular size of Ganymede was 1.6 arcseconds, and features only as small as 500 kilometers can be seen.

The inadequacy of ground-based observations was replaced by the first on-site spacecraft views provided by the Mariner, Voyager, Pioneer, and Cassini missons, which flew-by the planets starting in 1964, and later such missions as Magellan (Venus) MESSENGER (Mercury), Mars Reconnaissance Orbiter (Mars), as well as Dawn

(Ceres, Vesta), New Horizons (Pluto), and a number of asteroid and comet studies by Giotto (Halley's), Deep Space 1 (Comet Borelli), NEAR (Eros and Mathilde), and Rosetta(Comet 67P/Churyumov-Gerasimenko and Asteroid Lutetia). The list of minor bodies in the solar system for which we now have kilometer-scale images or better is an impressive one as the table shows.

Name	Mission	Diameter	Resolution
Moon	LRO	3,475 km	0.5 meters
Mercury	MESSENGER	4,878 km	10 meters
Venus	Magellan-Radar	12,100 km	100 meters
Mars	MRO	6,779 km	1 meter
Mars	Curiosity rover		millimeters
Phobos	MRO/HiRISE	22 km	20 meters
Deimos	MRO/HiRISE	12 km	20 meters
Ceres	Dawn	950 km	35 meters
Vesta	Dawn	525 km	70 meters
Comet 67P	Rosetta	4 km	0.1 meters
Comet 67P	Philae lander		0.05 meters
Halley's	Giotto	15 km	100 meters
Temple 1	Stardust	8 km	20 meters
Hartley 2	Deep Impact	1.5 km	7 meters
Wild 2	Stardust	5 km	20 meters
Borrelly	Deep Space 1	8 km	45 meters
Eros	NEAR	34 km x11 km	20 meters
Eros	NEAR-landing		0.1 meters
Mathilde	NEAR	50 km	180 meters
Gaspra	Galileo	19 x12 km	54 meters
Ida/Dactyl	Galileo	16 km	30 meters
Io	Galileo	3,636 km	9 meters
Europa	Galileo	3,160 km	9 meters
Ganymede	Galileo	5,262 km	74 meters
Callisto	Galileo	4,800 km	20 meters

Name	Mission	Diameter	Resolution
Ephimetheus	Cassini	129 km	224 meters
Janus	Cassini	203 km	448 meters
Mimas	Cassini	396 km	240 meters
Enceladus	Cassini	500 km	176 meters
Tethys	Cassini	1076 km	370 meters
Dione	Cassini	1128 km	23 meters
Helene	Cassini	43 km	187 meters
Rhea	Cassini	1532 km	460 meters
Titan	Cassini radar	5,150 km	300 meters
Titan	Huygens lander		20 meters
Hyperion	Cassini	270 km	362 meters
Iapetus	Cassini	1,492 km	400 meters
Phoebe	Cassini	213 km	74 meters
Miranda	Voyager 2	471 km	662 meters
Ariel	Voyager 2	1,157 km	2000 meters
Umbriel	Voyager 2	1,169 km	10000 meters
Titania	Voyager 2	1,576 km	13000 meters
Oberon	Voyager 2	1,522 km	12000 meters
Triton	Voyager 2	2,700 km	600 meters
Pluto	Hubble	2,360 km	800 kilometers
Pluto	New Horizons		25 meters

MRO=Mars Reconnaissance Orbiter; LRO=Lunar Reconnaissance Orbiter

The total mapped area represented by this table is 285 million square kilometers, which is about twice the land area of Earth. Of this total, 82 million km² or an area about equal to three times the area of Africa, has been mapped at better than 10-meters resolution. An area comparable to Africa has been mapped between 10 and 100 meters resolution. An area equal to the entire land mass of Earth has been mapped between 1 km and 100-meters resolution. Finally, an area equal in size to Australia has been mapped at a resolution between 1-10 kilometers. What this means is that all of the major planets and moons have now been mapped at better than 1 km resolution, so we know the principle land forms, craters, mountains and valleys. Moreover for Mercury, Mars, and the moon, we have 1-10 meter resolution maps allowing us to see boulders and changing geographic landforms such as the appearance of small impact craters. For five objects: the moon, Venus, Mars, Eros, Comet 67P and Titan, we have successfully deployed landers, and in one case (the moon) have deployed teams of astronauts to give first-hand reports and collect over 800 kg of samples. The surfaces of these objects are known to resolutions of centimeters and millimeters at least at some locations.

(Mars surface at millimeter resolution! Image credit: NASA)

This international process of mapping all of the major solar system non-planetary bodies at high resolution continues with a suite of future missions to asteroid 1999 JU3 (Hayabusha 2 and surface rover), Juno (Jupiter), ESA-Europa (Jupiter), and that takes us to the end of the 2020s. The current hot spots in the outer solar system are Europa and Enceladus because they seem to have active subsurface liquid oceans, and are intriguing places for searches for life. Beyond these moons, there has been no significant scientific interest in supporting very expensive billion-dollar missions to the outer solar system when the only apparent purpose is to get higher-resolution images of icy moons. We have largely passed the hey-day of planetary surveys, with all of the significant bodies of interest having been mapped at sufficient resolution to quantitatively understand their history and current topology. Future missions will focus on the handful of large moons with signs of subsurface water (Europa and Enceladus) and interesting surface activity (Titan, Pluto). Attempts at drilling into them to reach the liquid zone, and sample return missions are likely to follow in the 2030s and beyond, but more distant destinations generally have little to recommend them.

Once the Cassini mission ends in 2017 and the Juno mission completes its 33-orbit, radiation belt studies of Jupiter also in 2017, we may have one more outer solar system scientific opportunity when New Horizons spacecraft flys-by a Kuiper Belt Object called PT1 in 2019, but then that will be the end of the current era of research in the outer solar system. The remaining activity will focus on sample return missions from comets, asteroids, and more rovers on Mars with sample-return capability. There continues to be some interest in lunar exploration as well, especially in the south polar region where water-ice deposits appear to be present. Similar deposits have also been detected on Mercury.

In addition to the major planets, their moons and asteroids in the asteroid belt between the orbits of Mars and Jupiter, there is an additional population of significant bodies of interest to Earth, namely, the Near Earth Objects (NEOs). These are asteroidal bodies over one meter in size, whose orbits come 'very close' to Earth's orbit and often intersect it. Major operations are in progress to identify and catalog all of these objects believed to number in the 100s of thousands. Because they can approach Earth within the orbit of the moon, they are extremely interesting objects beyond their obvious potential for causing a direct impact on Earth and the devastation that would likely result. We might be able to capture and mine them!

The Minor Planets Center is the central reporting agency that holds the official sightings and orbital records for all objects in the solar system from meter-sized asteroids to the satellites of each planet. Currently, over 650,000 small bodies are known, and thousands of new ones are discovered every year. An example of the locations of some of these objects inside the orbit of Jupiter is seen in this figure. (Credit: Minor Planets Center)

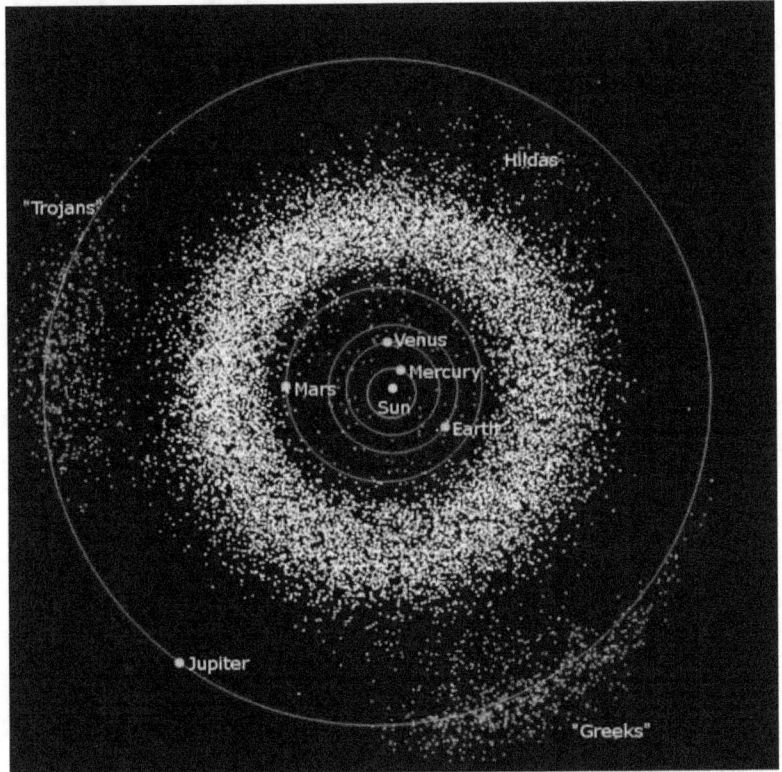

It is obvious that the inner edge just outside the orbit of Mars is not a barrier for asteroids to enter the inner solar system as far as the orbit of Mercury.

This astonishing map was calculated for all known objects on May 8, 2015. The clustering of objects near Earth is a selection effect because we discover more objects when they are closest to us. The orbits from top right to lower left are Mercury, Venus, Earth and Mars. (Image credit: Armagh Observatory)

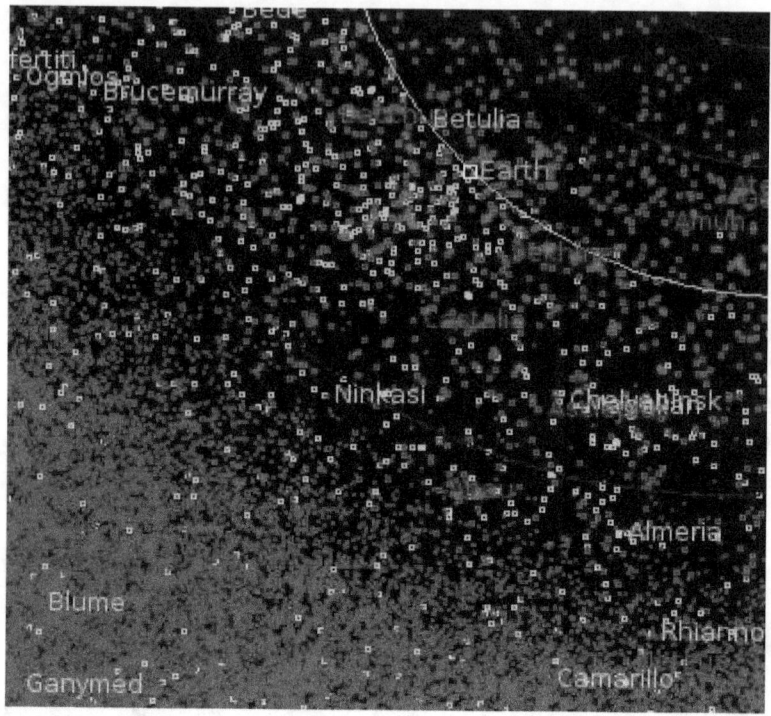

Asteroid orbits near Earth. (Credit: Armagh Observatory).

Another list of Near Earth Impact Hazards lists those 569 identified objects whose current orbits bring them very close to Earth, and based on the current orbit data and computation limits, may eventually have high collision probabilities. The vast majority of these objects, so far, are less than a few hundred meters across, but it is believed they represent only the vanguard of a much larger population of small bodies yet to be discovered.

On February 15, 2013 a previously undetected small body 70-meters across approached Earth from the day-side and exploded in the air over Chelyabinsk, Russia. It injured thousands of people from flying glass and building debris in an otherwise small town in Russia. It made international headlines as the new Tunguska Event. Ironically, at the same time, most people were focused on a NEO called 2012 DA14, which would be passing within 27,000 km of Earth's surface, and inside the geosynchronous satellite orbit. It was discovered the previous year with an estimated diameter of 45 meters; about ½ the size of the Chelyabinsk object. Unlike the Chelyabinsk object, which originated in the asteroid belt, DA14 was an object whose orbit was nearly circular and entirely between the orbit of Venus and Earth.

Chelyabinsk Meteor (Credit: GlobalNews.com)

Although Earth does exist within a veritable shooting gallery of interplanetary rocks and boulders, there is no reason why we can't use this misfortune to our advantage. By 2015, over 12,000 Near Earth Asteroids have been identified, and of these 1330 are of interest to NASA because they require less fuel and time to visit than current estimates for a journey to Mars. NASA is even planning the Asteroid Redirect Mission for the early 2020s to visit one of these small rocks a few tens of meters across, to capture it, and slowly drag it into a parking orbit around the moon! More about that later. But there are smaller objects too. This image is of the Willamette Meteorite, which is an iron-nickel meteorite with a mass of about 16 tons. On display at the American Museum of Natural History in New York City, it is classified as a type III iron meteorite, being composed of over 91% iron and 7.62% nickel, with traces of cobalt and phosphorus. The approximate dimensions of the meteorite are 10 feet (3.05 m) tall by 6.5 feet (1.98 m) wide by 4.25 feet (1.3 m) deep. The largest recovered meteorite in the world was found in 1920 in Namibia and has a mass of 64 tons. It is also an iron meteorite.

Why should we go?

When explorers on Earth traveled to other places it was often motivated by the quest for new resources, and the prospects for unusual adventure. Typical ventures were enabled by conventional technology and transportation such as sailing vessels, horse and mule transport, or simply walking. The major voyages, such as the journeys of Columbus, Magellan and Captain Cook, were expensive affairs that were supported by wealthy patrons, but hardly cost a significant portion of a county's wealth.

Our first contact with the concept of leaving Earth entirely and journeying to another celestial body, the moon, was not the undertaking of a few wealthy benefactors using conventional technology. It took a decade of hard work by more than 100,000 people, and in current dollars cost $100 billion. Entirely new technologies had to be invented, as did new techniques for communication, docking, maneuvering in space, and operating under an entirely new set of gravitational realities. The fundamental reason for this investment was political. The two nations that had the resources to accomplish this feat were locked into a pitched political battle to win-over new converts to their particular political theology: democracy vs communism, by demonstrating the technological and intellectual superiority of their respective politics. This allowed the United States to expend a miraculous 4% of its annual federal budget in 1968 to pursue this ideological war, despite that fact that the general public was lukewarm about the merits of space travel versus 'solving poverty here at home'. Once the Space Race had been won in the early-1970s, budgets for space exploration shrank from 4% to about 0.5% of the annual US budget almost over-night, and have remained there ever since.

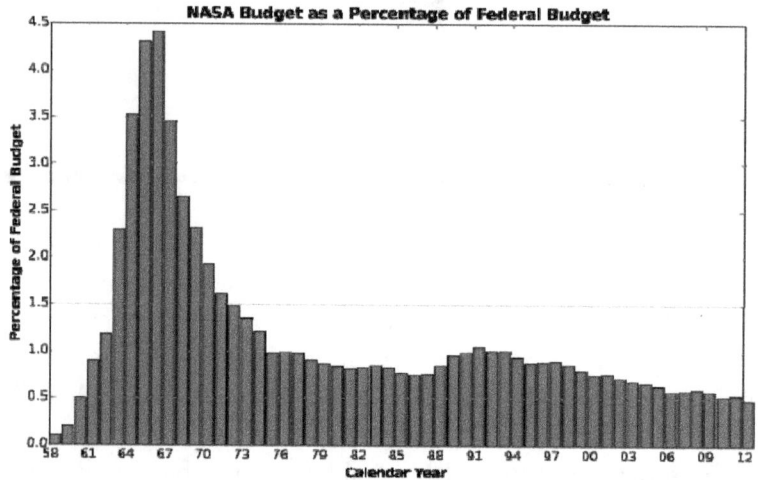

Since the 1970s, we have all become very fond of public surveys to gauge how well specific ideas or programs are doing. When you look back at the US public's opinion of space research since the 1960's you discover that citizens have always had significant problems with the space program. A Harris survey taken a year after the 1969 Apollo 11 moon landing revealed that half the public (56%) didn't think the landing was worth the money, but nearly everyone (81%) felt that nothing could equal actually seeing the astronauts walking on the moon. Year after year, the surveys show the bleak opinion that the US spends way too much on space exploration (78%). Yet despite not wanting to actually pay for the effort of traveling into space, 30% expect that we will have colonies on Mars by 2064. The percentage who think that the current level of funding is adequate or should be cut back even further amount to some 7 in 10 Pew Research survey respondents in 2010.

A recently released tape of a White House meeting that took place on November 21, 1962 between President Kennedy and NASA Administrator James E. Webb demonstrates this fact beyond all dispute. Kennedy explained, *"Everything that we do should be tied into getting on to the Moon ahead of the Russians. We ought to get it really clear that the policy ought to be that this is the top priority program of the agency and one of the top priorities of the United States government."* He added: *Otherwise we shouldn't be spending this kind of money, because I am not that interested in space. I think it's good. I think we ought to know about it. But we're talking about fantastic expenditures. We've wrecked our budget, and all these other domestic programs, and the only justification for it, in my opinion, is to do it in the time element I am asking.*

An independent assessment conducted by the National Research Council in 2012 noted that a crewed mission to Mars "has never received sufficient funding to advance beyond the rhetoric stage." Most experts advocate sustaining U.S. leadership in space. *"I'm convinced that in this century the nations that lead in the world are going to be those that create new knowledge. And one of the places where you have a huge opportunity to create new knowledge will be exploration of the universe, exploration of the solar system, and the building of technology that allows you to do that,"* said former congressman and aerospace expert Robert Walker at a CFR meeting on space policy in 2013.

In 2001, 48 percent of those surveyed by the National Science Foundation thought spending on space exploration was excessive; the highest percentage for any item in the survey—and nearly double the number of those who felt that the government was spending too much on national defense. Not since 1985 (before the Challenger accident), have more than 50 percent of respondents to NSF's public attitudes survey stated that the benefits of the space program exceeded the costs. NSF survey data suggest that most of the public is having difficulty recognizing the benefits of the space program. Those identified as attentive to science and technology, or space exploration are more likely than the public at large to

27

believe that the benefits exceed the costs. In 2001, at least 60 percent of each attentive group put the benefits ahead of the costs compared with less than 50 percent of the public at large.

In 2007, The Space Review conducted a survey in which respondents were asked what percentage of the national budget is allocated to NASA and to the Department of Defense. NASA's allocation, on average, was estimated to be approximately 24% of the federal budget. The Department of Defense was estimated on average to receive approximately 33%. In other words, a significant number of respondents believed NASA's budget approaches that of the Department of Defense, which in actuality receives almost 38 times more money. Once people were informed of the actual allocations, they were almost uniformly surprised. "Our favorite response came from one of the more vocal participants, who exclaimed, "No wonder we haven't gone anywhere!"

Meanwhile, Congress is clearly more knowledgeable about the value of space exploration than the voting public, and maintains our national effort year-after-year at about 0.6% of the federal budget. The Space Exploration, Development, and Settlement Act of 2015, drafted by Rep. Dana Rohrabacher (R-Calif.), would mark the second time in the last three decades that Congress has directed NASA to support efforts for permanent human settlements beyond Earth orbit. *"The expansion of permanent human presence beyond low-Earth orbit in a way that enables human settlement and a thriving space economy."* Under the bill, NASA would be required to produce a report every two years *"which describes the progress made toward expanding permanent human presence beyond low-Earth orbit"* in support of space settlement. The NASA Authorization Act of 1989 also included a provision requiring NASA to support space settlement activities. That section is commonly referred to as the Space Settlement Act of 1988, as it was originally introduced as a stand-alone bill by the late Rep. George Brown (D-Calif.).

Texas Senator Ted Cruz (R-Tx) isn't a proponent of big government programs, but this outspoken Tea Party Republican, while decrying the scientific evidence for climate change, championed NASA's human space exploration program as a national priority that deserves congressional support. Developing a rocket and capsule to take astronauts to Mars by the 2030s — a mission projected to cost tens of billions — is "critical" to ensuring the nation's leadership in space, according to Senator Cruz.

Based on their 2014 congressional meetings, *Citizens for Space Exploration* teams conducted 354 visits with congressmen and found that there was strong bipartisan support for NASA with more understanding of NASA's space exploration programs than in previous years. It was generally agreed that Mars is the ultimate destination for human space exploration, however, most agreed that the case for the value of human spaceflight will have to be made continuously, which means every year as though its past history and successes do not matter. It has to be justified based on what consumers like to ask *'What will it do for me right now?'*

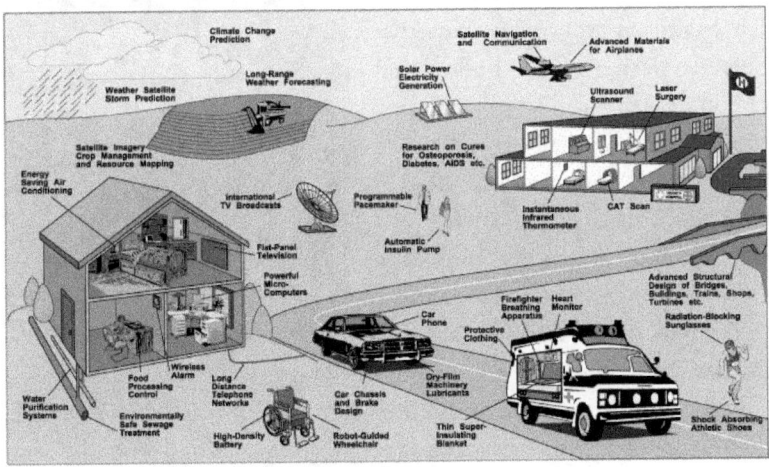

NASA spinoffs (Credit: NASA/Johnson SC)

Interplanetary Travel

Another popularly cited benefit of space exploration is job creation, or the obvious benefit that a space agency and its network of contractors, universities and other entities help people stay employed in high-tech careers. In 2012, NASA administrator Charles Bolden published a blog post about the Curiosity Mars rover landing, which was picked up by the White House website. *"It's also important to remember that the $2.5 billion investment made in this project was not spent on Mars, but right here on Earth, supporting more than 7,000 jobs in at least 31 states,"* he wrote.

When you add the labor costs to fabricate technology, mine metals and mix alloys, build integrated circuits, design and testbed new technologies, and do the hard engineering and scientific research to justify the cost, the actual labor cost for any government program is about 80%. In other words, when $2.5 billion is given to NASA to run a program, 80% of that money goes to salaries of real flesh-and-blood humans all the way down the supply chain to the miner working in the field to extract the raw materials. If NASA buys a specially-designed integrated circuit from Intel for $10,000 to run a critical function in a research satellite, about $8,000 of that purchase is for the salaries of all the engineers and technicians that helped design and build that circuit. That is why investing money in NASA for ANY of its programs is investing in the US workforce to the tune of 80% of the funding. When we build the Curiosity rover, a huge labor force in the tens of thousands is actually involved all the way down the supply chain for every component in the rover from its titanium skeleton and microscopic circuit

chips, to every single specially-designed screw and bolt. None of this money goes to Martians but stays right here in our paychecks!

In a 2005 paper, Roger Launius, chief historian at NASA, wrote, *"While there may be many myths about Apollo and spaceflight, the principal one is the story of a resolute nation moving outward into the unknown beyond Earth."* Nostalgia for the Space Age is rooted more in *The Jetsons* TV series than in reality. The myth of the "Sputnik moment" means that we spend time hand-wringing over a lack of shared ambition, rather than actually working toward game-changing goals. Time is wasted as we act like petulant children, whining that no one wants to go to Mars anymore, rather than making the case for a manned Red Planet mission.

A glimpse of the future. (Image Credit: Gunter Radtke)

You almost have to feel bad for the baby boomers (like me!) for not getting the future they were promised. When they were kids, there was a deliberate effort to get children excited about, and emotionally invested in, scientific and technological progress. Dr. Athelstan Spilhaus, the dean of the University of Minnesota's Institute of Technology, started one of at least two Sunday newspaper comic strips borne out of concern that American kids were falling behind the Russians intellectually and weren't sufficiently interested in science and technology. The comic explained scientific principles, often with a futuristic flair, and by 1959, Dr. Spilhaus' "Our New Age" appeared in more than 100 U.S. newspapers.

Compared to other space programs, some fear that NASA has lost its youthful exuberance. In 2014, China by some accounts had the most active and ambitious space program in the world. Chinese officials speak proudly of developing their lunar exploration program, building a heavy-lift rocket, constructing a spaceport, and planning an orbital space station. *"They are having launches, and in the United States we're in gridlock,"* says Joan Johnson-Freese, a professor at the U.S. Naval War College, in Newport, R.I. *"The Chinese will have a rover on the moon, and we're still developing PowerPoints for programs that don't get approved by Congress."*

When asked about his agency's biggest challenge, Ma Xingrui, director of the China National Space Administration, spoke of engineering complications. When Charles F. Bolden Jr., the NASA administrator, was asked the same question, he had quite a different answer. *"I think NASA's biggest challenge is inspiring our nation. We need to inspire the American public, and we need to inspire this Congress. Because that translates to funding."* China is where the US was when President Kennedy announced our moon program and only the political reason mattered as a matter of national defense. Today, the value of space exploration is completely decoupled from any similar life or death political concern. It is marketed as largely a jobs program for high-tech workers and scientists, who represent only a vanishingly small share of the domestic labor force. Your average person-on-the-street has never met a 'rocket scientist' or any other scientist working for NASA.

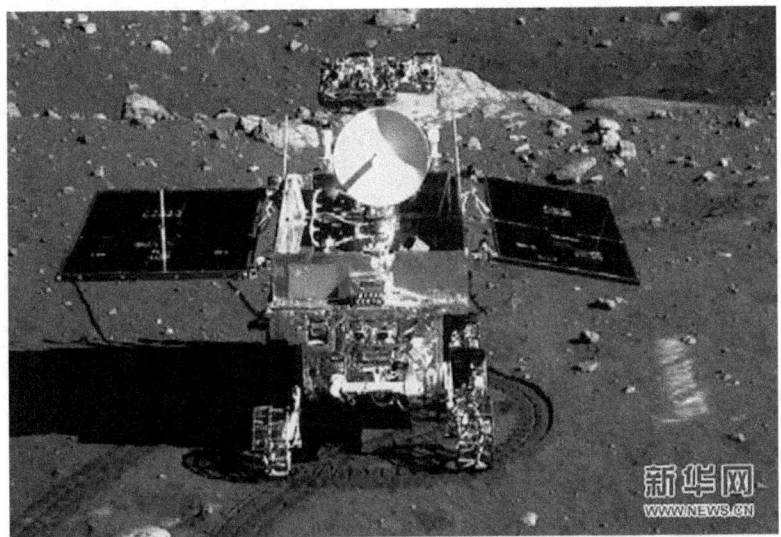

Chinese Yutu rover on the moon. (Credit: CNSA / CCTV)

In 2012, after the adoption and abandonment of the *Vision for Space Exploration* to take people to the moon and Mars, the National Academy of Science's National Research Council became the latest in a long line of expert panels to conclude *"There is no strong compelling national vision for the human spaceflight program, which is arguably the centerpiece of NASA's spectrum of mission areas...NASA needs to embrace a singular, unambiguous purpose that leverages its core strengths and provides a clear direction for prioritizing tasks and assigning resources."* But what does that mean when every good idea about exploring the moon and Mars, or deploying nuclear rocket technology is shot down as 'not having a goal'?

Deep-space exploration got a jump-start in 2004, when George W. Bush announced the Constellation program; a plan to return astronauts to the Moon's surface by 2020, and to Mars sometime afterward. The program, however, was consistently underfunded, putting it way behind schedule, and President Obama finally had to put it out of its misery by killing it in 2010. He couldn't kill it entirely, though. The Orion and Ares projects were already providing thousands of jobs at Florida's Kennedy Space Center and elsewhere. Lawmakers representing those districts fought hard to prevent the jobs from disappearing, and eventually directed NASA to build a new rocket system called the Space Launch System (SLS) to replace Ares. SLS, though, would need something to do. So President Obama proposed landing humans on an asteroid in the 2020s as a "stepping stone" for an eventual trip to Mars in the 2030s, retaining Orion as a manned expedition capsule. The Asteroid Redirect Mission (ARM) was later proposed as a cheaper unmanned probe that would be used to grab the asteroid and put it in orbit around the moon.

Even with private companies participating, the general perception is that NASA needs to do more to justify a Mars mission to Congress and the public. According to Marco Caceres, a space analyst with the Teal Group "*The biggest problem is that they haven't really been able to explain why we want to do this. Without a clearly*

articulated reason, we're just going to keep on running in to the same problems over and over."

According to Charles Walker, a former engineer and astronaut who flew on three space shuttle missions in the 1980s, *"In a recent letter cosigned with 28 other members of the House, Reps. Frederica S. Wilson, Corrine Brown, Bill Posey and Alan Grayson urged the White House to put a greater emphasis on efforts to send American astronauts to explore space beyond earth's orbit.*

NASA's deep-space efforts complement the flurry of activity by private companies to reach the International Space Station. As the first astronaut to fly with NASA on behalf of a commercial space company, I've seen how a smart division between NASA and the private sector can drive costs down for profitable missions, leaving NASA with more money to pursue greater space exploration.

Some argue that space exploration is a luxury we can't afford. But if we want to continue America's global economic leadership, I'd say it's a necessity, not a luxury. Investing in NASA generates enormous returns — upwards of $10 in lucrative spinoffs for each dollar spent in research and development — at very little cost. The entire NASA budget is less than half a penny out of each taxpayer dollar. And deep-space missions would not break our budget. To put them in perspective, we spend as much to maintain empty government buildings every two years as it would cost to build the SLS and Orion capsule needed to reach Mars."

NASA awarded $6.8 billion in contracts to Boeing and SpaceX in 2014 to finish developing their own space vehicles to carry U.S. astronauts to and from the International Space Station by 2017. Representative Dana Rohrabacher has even suggested that the US should outsource its Mars mission to a private contractor such as SpaceX. NASA Administrator Bolden countered that *"no commercial company is going to independently go to Mars without support of the government,"* and said that he believes relying on a private company would only slow the mission.

Meanwhile, the one goal that does seem to have some support continues to be the Asteroid Retrieval Mission to be launched by 2019. The program scientists have three asteroids in mind thus far: Itokawa, Bennu, and 2008 EV5. It would launch a robotic spacecraft to intercept and land on the rock before plucking off a good-side boulder and maneuvering this smaller object into a lunar orbit. The next-generation SLS and Orion capsule manned mission would then visit the ARM boulder shortly thereafter. Unfortunately, Congress hates the asteroid recovery idea.

Congress claims to support the Mars mission plan, but has been sharply critical of NASA's ARM project. In fact, Congressional representatives are so opposed to taking the SLS on any kind of test or proving mission that they attempted to pass a 2013 bill that would forbid funding for ARM altogether.

"A frequent objection raised against scenarios for the human settlement and terraforming of Mars is that while such projects may be technologically feasible, there is no possible way that they can be paid for. In 1802, Napoleon Bonaparte sold a third of what is now the United States for 2 million dollars. The existence of Australia was known to Europe for two hundred years before the first colony arrived, and no European power even bothered to claim the continent until 1830. These pieces of short-sighted statecraft, almost incomprehensible in their stupidity, are legendary today. Yet their consistency shows a persistent blind spot among policy making groups as to the true sources of wealth and power."

How about private industry? With all the excitement surrounding SpaceX, and the success of PayPal and Tesla Motors, one might expect Elon Musk to take SpaceX public. So why hasn't he? *"The reason I haven't taken SpaceX public is the goals of SpaceX are very long-term, which is to establish a city on Mars,"* In November of 2012, redOrbit reported about Musk´s designs for a Mars colony in the next 20 years.

"In his plans, Musk said [the initial Mars colonization effort] would start with a pioneering group of less than 10 people who would be transported through a reusable rocket powered by liquid oxygen and methane." This first group would literally build a sustainable colony from the ground up, including housing, crops, buildings and everything needed to sustain life on Mars. Eventually, Musk said he would like to have 80,000 colonists by the 20-year mark. Of course, at an estimated price of about $500,000, a one-way ticket to the Red Planet is still pretty steep.

Another group interested in colonizing Mars is a nonprofit organization from the Netherlands that is planning a settlement by 2023. Dutch entrepreneur Bas Lansdorp is leading a group of his countrymen putting together its plans for the *Mars One* settlement. *"Mars One will establish the first human settlement on Mars in 2023,"* the group said in a statement on its website. *"A habitable settlement will be waiting for the settlers when they land."* Then every two years thereafter, a new crew would arrive on the Mars to replace the previous occupants.

Both of these groups are working feverishly to find funding, participants and suppliers. However, these are not the only hurdles that they face. In the end, it all comes down to public interest.

The confused state of public opinion.

Many surveys have touted the long term result that nearly 25% of the public want to see the US space program eliminated, and that only 56% see any value in the long term benefits of space exploration. Although these numbers seem very bad in absolute terms, they should be compared with other statistical opinions revealed from these kinds of surveys, which basically tell us that what the public believes is quite irrelevant to how Congress funds basic research. Voters elect politician for their position on abortion, terrorism, jobs and gay rights. The issue of how we should fund science and NASA never appears in public campaigns.

Vaccinations: This is a powerful tool that has all but eradicated polio, measles and rubella since the 1950s, yet the value of vaccinations plus popular misinformation about side effects now threatens a new era of disease resurgence. According to a Pew Research survey in 2015, over 20% of adults over age 50 say that parents should get to choose whether children get vaccinated, while 41% between ages 18 and 29 also favor parental rights. Whether you get to choose to not get vaccinated and then become a life hazard for the child next door with a compromised immune system should never be 'your right to choose'.

Climate change: Most Americans 'believe' in climate change but give it a low priority in terms of risk and impact to themselves. Although 61% said there was solid evidence for climate change, only 41% considered it a major priority compared to the spread of ISIS and the nuclear programs of Iran and North Korea.

Seat belts – In a 1992 report to Congress, the annual statistics showed that over 15,000 lives could be saved every year if front-seat passengers wore their seatbelts. By 1991, 41 states had mandated seatbelt laws. Nevertheless, these laws are historically highly unpopular because citizens feel it infringes on their freedom to choose.

Social Security and Medicare – When it was first enacted by President Roosevelt in 1935, about 25% of the public did not favor it at all for a variety of reasons including the 'taxation' issue and its obvious socialistic origins. There was a campaign in 1961, organized by the American Medical Association called Operation Coffee Cup, and it had as its spokesman, a well-known actor, Ronald Reagan, opposing a law guaranteeing health insurance for the elderly. Support for the program is currently at 80 %, but that still means that 20% of the public do not want this federal program at all!

Gasoline Tax - The total US average fuel tax is about 48 cents per gallon for gas and 54 cents per gallon for diesel. Gallup tested the proposal of a 20 cent per gallon tax being proposed by many states to fund infrastructure and mass-transit projects. 66% of surveyed voters opposed the proposal. Pothole damage, meanwhile, costs commuters about $6.4 billion a year. When you consider that commuters use about 136 billion gallons per year, that works out to 5-cents per gallon 'tax' for potholes alone. *"People may feel the financial impact of underinvestment most directly on the roads. Researchers at the Texas A&M Transportation Institute estimate that, unless spending increases, congestion and rough roads will cost the average Texas household $6,100 a year in wasted fuel, vehicle repairs, and time lost sitting in traffic between now and 2035."* In 20 years, that works out to a 'tax' of $300 per year per household. Apparently a significant number of Americans prefer to pay this form of tax rather than the 54-cent gas tax. This is why you never trust an average American to get any aspect of federal budget investments correct.

Interplanetary Travel

US Interstate Highway System –The Federal Highway Act of 1938, and later the Federal Aid Highway Act of 1956, created 47,856 miles of paved roadways in the United States. About one-quarter of all vehicle miles driven use the Interstate system. The cost of construction has been estimated at $425 billion (in 2006 dollars). In 1956, Congress inaugurated the Interstate Highway system not as a public service but as a matter of national defense so that the US military had speedy access to all locations in the 48 states on a consistent set of roadways, bridges and tunnel systems. Gallop Poll surveys at that time showed that 76% of the public agreed that "*more express highways are needed between the large cities*" of the US. That means 24% did not think such a massive investment in roadways was needed. In the Rockefeller Foundation Transportation Survey, more than 1,000 US voters were surveyed by Hart Research Associates and Public Opinion Strategies in 2011. The survey found overwhelming support for increased transportation investment, but hardly any support for increasing taxes or the use of tolls. About 93% felt that improving the country's transportation infrastructure was important. But 71% said it would be unacceptable to increase the federal gas tax, and 64% were against adding more tolls to highways and bridges.

These are examples of programs that everyone benefits from even if they don't like the program or how the particular benefit is funded. Most of the time we don't even know how much we are paying for a benefit, and usually as for space exploration, we think these unknown costs are huge and unwarranted. Although everyone can tell you how much they pay for each of their utilities, cell phones and cable TV each month, few realize that the entire US roadway system costs us about $18 per month. They also don't realize that NASA's entire budget amounts to about $4 per month per taxpayer. That is about the cost of one Starbucks Grande Hot Chocolate!

There are also examples of programs that only a few benefit from directly, but the general population benefits from as well even if they don't like the program.

The Affordable Healthcare Act is a largely Republican program enacted by a Democratic president, which provided health insurance to nearly 20 million people who previously could not afford health insurance. Typically, they would wait until their health situation became critical, present themselves to hospital emergency wards for free care, and the costs were passed on to everyone else through the hospital cost structure and insurance coverages and co-pays. By 2015, 43% of the surveyed public had an unfavorable view of this program (30% democrats and 75% republicans). Nevertheless, by the fifth year, over 16 million people had insurance who didn't have it before. The Congressional Budget Office reports that the rate of increase in real health spending per person is at its lowest point in 50 years and more than 3 percentage points under the historical average.

Even as parents were buying space toys for their kids and encouraging them to read the more educational Sunday funnies, they were skeptical, even surly, about the funds spent on Apollo and NASA. Those people with reservations about the space program seem to be primarily concerned that the money spent on space could be better invested in more earthly problems. *Why should we spend money in space when we need it here on Earth?"* But what does this concern really mean? This is actually a bad misunderstanding of just where money for space exploration actually goes. The construction of expensive satellites and other resources is not done by Martian laborers. It is done by humans working in a variety of settings from industrial venues to fabricate the launch vehicles, to university laboratories to fabricate high-tech instruments. It is done by thousands of people working at major companies like Lockheed, General Dynamics, SAIC, Boeing and Rocketdyne, or by small teams of two to ten people working at universities, or even small 'mom and pop' companies all over the country. For example, the Apollo spacesuit was created by the Playtex Corporation, who famously also created women's undergarments. About 15 people were involved in sewing a single suit, and dozens of suits were required for each Apollo mission. (Image credit: International Latex Corporation (Playtex) Dover, LP)

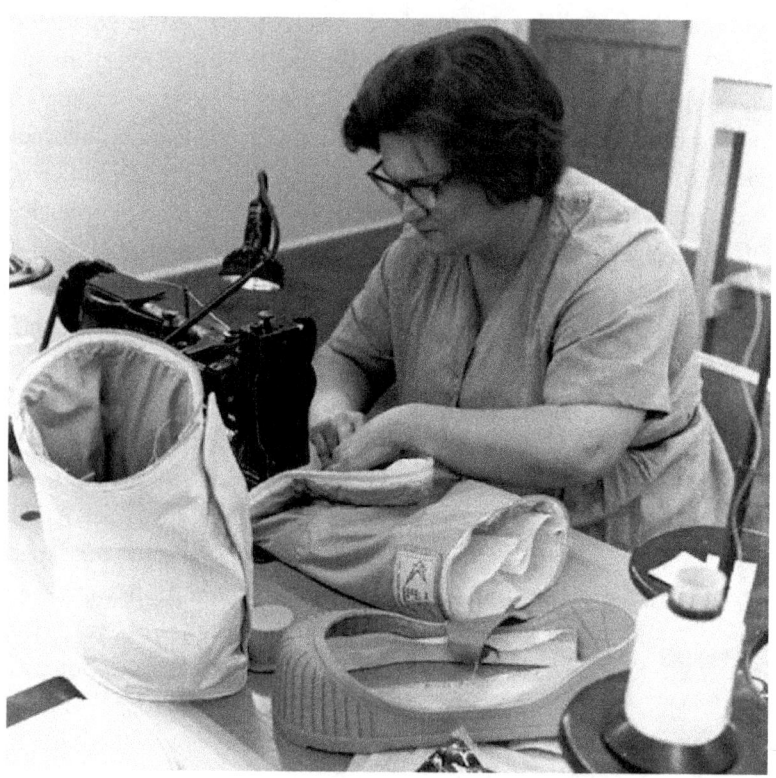

How much will it cost?

In 2014, NASA contracted with 257 US Small Businesses and 29 institutions for $48 million of work to explore technologies for future missions through their Small Business Innovation Research (SBIR) program. Technologies funded by these NASA innovation programs may one day find their way into journeys across the solar system, or into your iPhones, laptops and entertainment systems among other places. NASA is funding proposals to enable in-space transportation for human and robotic missions; new ways to keep astronauts safe on their journey, and innovative ways to keep spacecraft systems fully operational.

Take for example *Litespeed Bicycles*. This Chattanooga, Tennessee-based bicycle company with a family of seven people, got involved with the Mars Curiosity mission when a NASA engineer working on the rover's titanium fabrication was also a cyclist familiar with *Litespeed*'s highly durable titanium bikes. He suggested NASA work with the company on the rover's titanium suspension arms. The arms, according to NASA, have the strength and precision to maneuver the 73-pound turret at the end of the arm accurately enough to deliver an aspirin tablet into a thimble.

The bottom line is that about 80% of any NASA program is labor costs, whether you are talking about fabricating a zipper for a spacesuit, or building the International Space Station. This is an amount of (federal, tax-payer) money that definitely stays here on Earth to help employ people who then have to pay mortgages, buy groceries and all the other things that constitute American Life. The real issue is whether you see money going to a workforce supporting the peaceful uses of space exploration and the development of more high-tech gadgets is of less national value than those building the Global Hawk ($2.5 billion) fifteen EA-18G Growler electronic warfare planes ($1.5 billion) or especially the F-35 Joint Strike Fighter ($400 billion), which is unable to view the battlefield and is by some estimates a ten-year step back in technology.

The F-35 is a massively expensive program that is fraught with severe technical problems but because it has the support of major defense contractors like Lockheed, the lobbying ability to keep this program going and supported by Congress is enormous. The F-35 funnels business to a global network of contractors that includes Northrop Grumman Corp. and Kongsberg Gruppen ASA of Norway. It counts 1,300 suppliers in 45 states supporting 133,000 jobs -- and more in nine other countries, according to Lockheed. The F-35 is an example of how large weapons programs can plow ahead amid questions about their strategic necessity and their failure to arrive on time and on budget. "It's got a lot of political protection," said Winslow Wheeler, a director at the Project on Government Oversight's Center for Defense Information in Washington. "In that environment, very, very few members of Congress are willing to say this is an unaffordable dog and we need to get rid of it."

So whether the public wants the space program or not is irrelevant. The bottom line is whether major private contractors with huge political clout can be persuaded to get on board and bid on space-related contracts. So far, the only significant contracts to come out of NASA that interest major companies are the ISS, the Shuttle and the Space Launch System. These are all supported by the NASA manned space program which gets $12 billion out of the $17.5 billion (FY 2015) annual NASA budget. These issues of support are settled, not by poor voter surveys about the necessity of space exploration, but by enough Congressional support for the budget requests by NASA for new and existing programs. Congress votes on the basis of how funding a particular program will create or maintain jobs for the constituents they represent. By making sure that NASA provides jobs for workers and businesses across all 50 states, NASA guarantees that it has a broad Congressional constituency that will continue to support its efforts even if the average voter is completely clueless about the value of space exploration.

The NASA budget process

First, NASA creates an annual budget to support its ongoing programs and any new programs it wants to include. This starts in March with a request by NASA's Office of the Chief Financial Officer to the various internal NASA Centers. In June this internal budget, plus four additional forecast years is discussed and finalized as the NASA budget request. This budget is then sent to the Office of Management and Budgets in September for analysis. By mid-January the finalized budget is sent to the Government Printing Office to run off copies for Congress and the President. The President then includes this in his formal request to Congress.

Then a lengthy period of discussion begins in the House and Senate through their special committees. The House reviews the President's NASA budget request, looks at every line in the budget for old and new programs, holds hearings, and comes up with its own budget request for NASA. The House then writes a NASA Reauthorization Act for that year and votes on it. For 2015 this bill was called HR 810 the 'National Aeronautics and Space

Interplanetary Travel

Administration Authorization Act of 2015'. The previous 2014 Act called HR 4412 was passed 401 to 2.

For the 2015 budget, ranking Member of the Space Subcommittee Donna F. Edwards (D-MD) said in her floor statement, *"NASA is a crown jewel of our Federal government. NASA's space and aeronautics programs help maintain our competitiveness, serve as a catalyst for innovation and economic growth, and inspire the next generation to dream big and garner the skills to turn those dreams into action. NASA and our space program have a long history of bipartisan support. NASA needs our constancy of purpose and direction now, and this one-year bill does just that while also allowing us the time needed to build on this baseline as we work toward a multi-year reauthorization over the coming year, once H.R. 810 is enacted into law."*

Meanwhile the Senate creates its own Reauthorization bill. In 2014 this was called S.1317. We now have four different budgets for NASA: The one requested by NASA and sent to the president; The one the President decides to include in his official budget request to Congress; The House Reauthorization bill (e.g. HR.4412 for 2014); and the Senate's version of the budget (e.g. S.1317 for 2014).

The last step is that the House and Senate have to negotiate and pass a final NASA budget, which is sent to the President to sign. It is Congress after all that actually writes and appropriates the funds. The Presidential budget is only his recommendation, and he will only see the NASA budget as part of a bigger political decision about whether to pass or reject the Congressional budget.

At each step of the Congressional budget process, members can be lobbied by 'friends of NASA' to support NASA's original recommendations, or even increase them. This lobbying effort includes discussions of jobs likely to be gained or lost in various congressional districts, or other factors. Generally it is a huge up-hill battle to significantly increase NASA's budget, but recent issues like the retirement of the Space Shuttle, and our loss of domestic access to space, have been potent motivators to support the creation of the new and expensive Space Launch System, and various commercial ventures with SpaceX and United Launch Services among others.

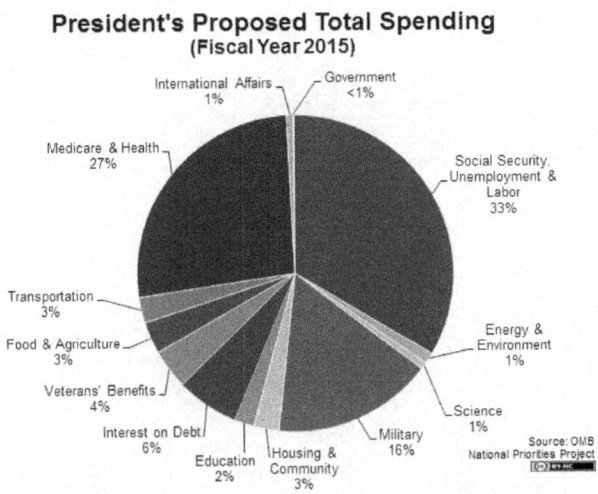

Here is the federal budget proposed by President Obama for 2015. All federal spending on non-military research and development (about $140 billion in 2015) fits into the relatively small "discretionary spending" portion of the federal budget. Discretionary spending, which is the only part of the budget that Congress and the White House can control from year to year, accounted for just 29% of the nearly $4 trillion the government will spend this year. The rest—71%—goes to so-called mandatory programs, such as the Medicare health program for senior citizens, and interest on the federal debt.

Six agencies account for more than 95% of total federal expenditures on basic research, which amounts to about $32 billion this year: NIH gets 50%, NASA gets 10%, NSF gets 14% and

Department of Energy gets 13%. Once NASA gets its budget, it still has to parcel this largess out to its many ongoing programs. All of the science missions you hear about amount to $5 billion in NASA's roughly $18 billion annual budget. The manned space program (ISS plus the development of the Space Launch System) takes up $8 billion.

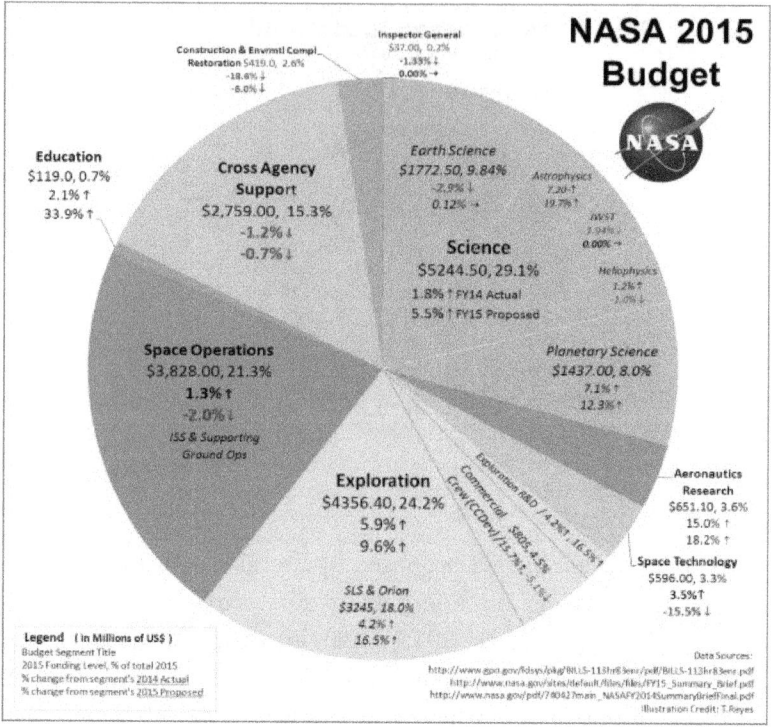

Congress repeatedly votes *en masse* to continue funding NASA at at least these levels because, unlike the general public, they see that NASA is a significant return on investment. A report by the Space Foundation estimated that space-related activities contributed $180 billion to the economy in 2005. More than 60% of this came from commercial goods and services created by companies related to space technology. This means that each dollar of NASA spending is a catalyst for $10 of economic benefit.

But all is not a bed of roses within the halls of Congress. NASA appropriations are complicated by a rift between the White House and Capitol Hill along party lines. The Democrat-controlled White House has favored robotics and the use of private enterprise to advance NASA, while Republicans on the Hill have supported the big human spaceflight projects backed by large aerospace industries (Boeing, Lockheed etc) who have powerful lobbies.

The other problem is that the NASA budget ($18 billion in 2015) is huge compared to other programs that have a larger social impact. For example, the entire budget for the National Cancer Institute is about $5 billion per year. The National School Lunch Program costs about $11 billion. The Head Start Program received $8.6 billion in 2014. When seen against all of these pressing needs, it is easy to decide that space exploration is a luxury we can ill afford. But of course that is the naïve way to look at space funding, and fortunately Congress does not pay much attention to public opinion when it comes to funding NASA. Congress DOES however weigh every increase in NASA funding and new mission starts against having to find this money in some other corner of the federal budget in the zero-sum, ideological climate of No New Taxes. Repeated arguments that NASA keeps high-tech scientists and STEM workers employed, and new inspired children coming into the career pipeline, only go so far. With No New Taxes as one popular mantra, and a steadily growing share of the budget going towards mandatory spending (Social Security, Medicare and debt payments), NASA is caught between a rock and a hard place to grow its budget beyond the current 0.6% of the federal budget. Nonetheless, NASA remains the largest space agency on the planet because of its huge ties to large commercial aerospace corporations who carry enormous political and lobbying clout.

The issue of increasing the NASA budget to include a more aggressive support of interplanetary travel is basically dead in the water. The value of NASA's many proposed ideas to capture an

The World Trails NASA in Space Exploration Expenditure
Annual budgets of international space agencies in 2013

NASA		$16.6bn
Roscosmos		$5.60bn
ESA		$5.50bn
CNES		$2.69bn
JAXA		$2.03bn
ASI		$1.80bn*
CNSA		$1.30bn**
DLR		$1.10bn
ISRO		$1.10bn
UKSA		$519.0m

*2014
**estimate

@StatistaCharts   Sources: Respective space agencies

statista

asteroid, return to the moon, or travel to Mars are all intensely controversial. The public input to these decisions is minimal and basically irrelevant. Even when NASA provides high-tech jobs and a 10 to 1 return on public investment, the public is shackled by a massive misunderstanding on the actual cost of NASA (many seem to think that 24% of the federal budget goes to NASA!) and Congress realizes that there is far more value to NASA as a federal enterprise, and will never pull its budget. The question is what circumstances will cause Congress and the President to, say, double the NASA budget so other projects can be attempted?

No one can honestly think of one.

Meanwhile, Congress is looking for a compelling reason why, within the current NASA budget, we should wrangle an asteroid, and return to the moon or Mars. NASA, and just about anyone that reads science fiction, can provide many good reasons. But Congress is looking for something more, which means they are particularly interested in reasons that make political sense. This takes us back to President Kennedy's reasons for focusing our national will on going to the moon. It was purely a political reason to out-compete the USSR. A similar justification today must also involve politics, not just the benefit of more high-tech jobs, or the working out of some human destiny to explore and expand. Other than China's rapidly advancing space efforts, we really have no reasons that make good political sense. Still, even our existing space program continues to have overwhelming bi-partisan support. Every President since Kennedy has lauded NASA for its contribution to our national advancement, for stimulating and inspiring our children, and making possible the major scientific discoveries along the way. The following chart shows the voting for H.R. 4412 the NASA Reauthorization Act of 2014 where 183 democrats (out of 199, and 218 republicans out of 232 approved the funding for NASA in 2014. Only 2 republicans voted not to fund and 28 abstained.

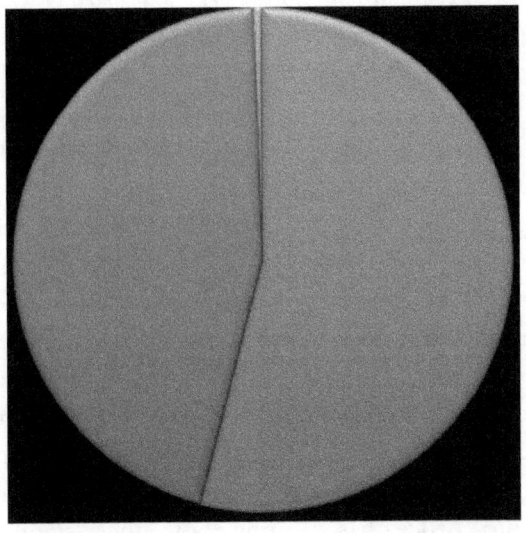

NASA Administrator Michael Griffin noted in an interview with Air and Space Magazine that *"I am convinced that if NASA were to disappear tomorrow, if we never put up another Hubble Space Telescope, never*

put another human being in space, people in this country would be profoundly distraught. Americans would feel that we had lost something that matters, that our best days were behind us, and they would feel themselves somehow diminished. Yet I think most would be unable to say why....As for national security, what is the value to the United States of being involved in enterprises which lift up human hearts everywhere? What is the value to the United States of being a leader in such efforts, in projects in which every technologically capable nation wants to take part? The greatest strategy for national security, more effective than having better guns and bombs than everyone else, is being a nation that does the kinds of things that make others want to do them with us.

What do you have to do, how do you have to behave, to do space projects? You have to value hard work. You have to live by excellence, or die from the lack of it. You have to understand and practice both leadership and followership. You have to build partnerships; leaders need partners and allies, as well as followers. You have to accept the challenge of the unknown, knowing that you might fail, and to do so not without fear but with mastery of fear and a determination to go anyway. You have to defer gratification because we work on things that not all of us will live to see—and we know it."

Detractors note that, once the US military completed their establishment of operations in near-Earth space, further interest in a larger (e.g. political) role for space ended, which is why there is no impetus to go to the moon and beyond. The motivations now have to come from the public/commercial sector. We do not need space for new resources or new territory to occupy, because these costs will be enormous and only involve small numbers of people much as Antarctic outposts do today. There is no economic reason to expand beyond geosynchronous earth orbit. There is no political or military reason either, and the general public will not vote a politician into office based upon how they stand on space exploration.

This, then, is the real reason why the moon was abandoned in 1972, and why instead of the 21st Century looking like a scene out of *2001:A Space Odyssey*, it more closely resembles a computer-generated, comfortable, but introverted, scene out of the movie, *The Matrix*. But it really doesn't have to be this way!

The USA is insanely wealthy. Everyone grouses about how hard it is to make mortgage and car payments, but when you consider how much money actually flows through the US economy every year it is nothing short of astonishing. Like so many other things in life, it is not that the resources are not available, it's that the priorities we assign to various efforts need to be modified. This is akin to the problem of fresh water availability that western states like California are now experiencing. It's not that fresh water is scarce (though arguably it is), it's that the way it is allocated is 100 years out of date and is based on antiquated ways of prioritizing how much farmers and cities get to share.

The USA had a Gross Domestic Product of $17.7 trillion during the first quarter of 2015, making it not only the wealthiest country in the world with 22% of the world GDP, but more wealthy than the next three countries, China, Japan and Germany, combined.

Another way to look at the GDP is by the four sectors that enter into its calculation: Consumption, Investment, Government Spending and the Balance of Trade (Exports minus Imports). For 2010, Consumption (Goods plus Services) was $10.4 trillion, Investment was $1.7 trillion, Government Spending was $2.9

trillion and the Balance of Trade was -0.5 trillion for a net GDP of $14.6 trillion in the first quarter of 2010.

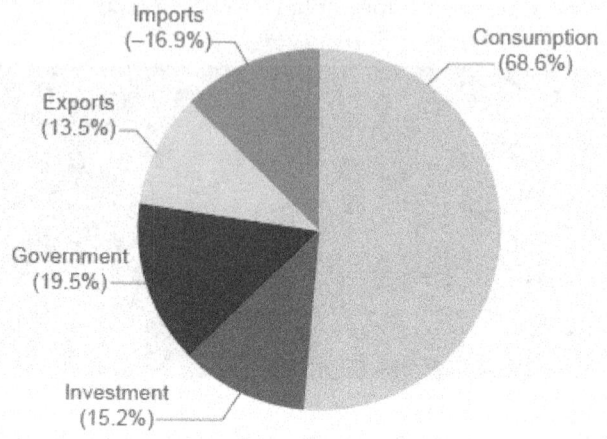

This figure shows the US GDP for 2012, which was $16.2 trillion. By 2014, about 25% of our annual GDP was for taxes which was $17.7 trillion. By the way, this makes us the 32nd lowest-taxed country in the world, with Denmark having the highest at 47%. Next we have consumer purchases at 70%. This is why we are considered a consumer-based economy. This works out to $12 trillion out of the nearly $18 trillion GDP in 2014 of which $4 trillion is goods and $8 trillion is services. Within the goods category we have durable goods (cars, furniture) at $1.3 trillion and non-durable (food, clothing, fuel) amounts to $2.7 trillion. The service sector produces most of the value in our economy. This does not imply that manufacturing will eventually disappear, but that more of the economic activity will be in the service sector. The service sector includes a mixture of professions that provide some kind of salaried service such as lawyers, sales staff, research scientists and hotel staff. Because twice our Consumer GDP is in services rather than goods, this is why some people complain that this country doesn't actually make anything anymore (Goods), we just sell each other French fries and clothes, and complain to our lawyers and bankers (Services).

When you buy a bottle of water, that adds to Goods. When you pay your monthly mortgage, that does nothing to the GDP except to take your money out of circulation that could have gone to increasing the Consumption portion of the GDP. That's why the US GDP has been stagnant since 2009 because people are paying down debt instead of buying more goods and services. Because the government and business are mesmerized by an ever-increasing GDP, they are actually not interested in you paying down your personal debt. They want you to contribute your earnings to one of the five GDP sectors!

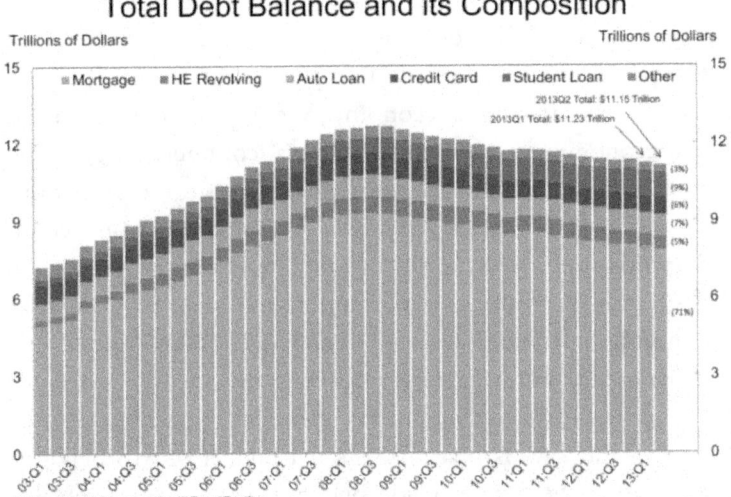

When you look at the long term trend for the US GDP it is relentlessly increasing. Have a look at this historical plot from 1970 to 2012.

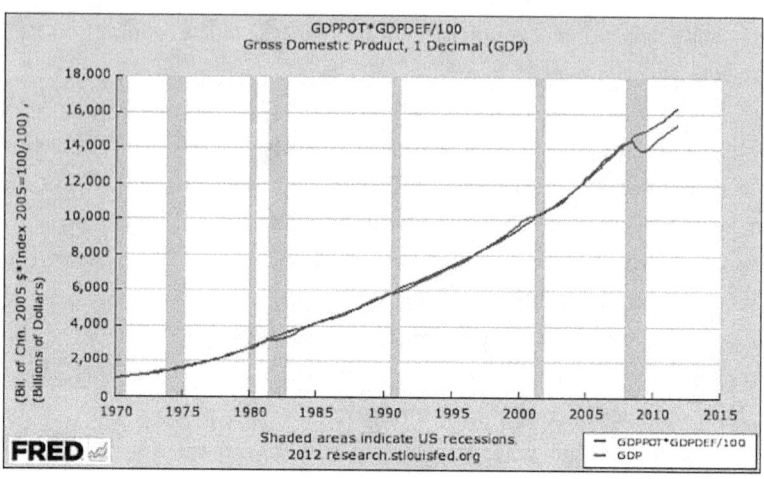

The US Great Recession of 2008 is clearly visible in the lower 'actual' curve, while the upper curve is where the economy should have been at its full productivity. The difference is $890 billion, ($7,800 per household) which has permanently been lost each subsequent year from our economy. We appear to be in a New Normal economic state that, nevertheless, continues to grow each year by the same amount as before 2008: about 3%. Economists look at this and say that unsustainable consumer debt financed the GDP growth and when 2008 hit and the financial markets for housing crashed, we have now returned to a rate of normal and more healthy growth that more properly reflects the level of consumer credit appropriate for our economy.

What all of this says is that NASA's flat budget of $18 billion per year (as well as the flat $5 billion budget for cancer research) is an embarrassing pittance compared to the volume of money that flows through our economy. It is not that cancer research or space exploration are costly undertakings, it is that so long as these are only supported by federal tax investment through research grants and federal contracts, they will always remain a part of the 25% of the GDP that currently goes to Government Spending, and not the much larger Consumer sector where the true wealth of this country lies.

The difficulty is that the Consumer sector is not about buying rockets, and scientists are not working for consumers as part of a Service industry. These functions are so removed from the Consumer sector that only Government Spending can really support them the same way that it supports building tanks, aircraft carriers and stealth bombers. This is why NASA is trying to engage commercial 'for profit' companies like SpaceX. Although in part these companies are on the Government Spending dole to deliver payloads to the ISS, they also have the opportunity to run a separate consumer-oriented business by literally selling tickets on their rockets. They can also sell their services to satellite manufacturers as a launch service. The question is whether this growing avenue of conducting space research will blossom into a major interplanetary commerce situation using laser communication and nuclear rockets. The answer will depend on how well a private company can find something in space to sell here on the ground. Will it be resources (mining)? Will it be entertainment (passenger tickets to a space station)? Will, it be scientific research (paid through private or government grants), or will it be some other attraction that can be sold to the consumer?

Long range budgets are very hard to predict because every year we have the option of changing the details of how we fund various programs and borrow to finance them. Here is one forecast by the Congressional Budget Office based on the government spending allocations and trends from ca 2008.

As many experts have been telling us, the policy of not raising taxes (the horizontal line at 18%) to pay for all of the mandated programs that Congress has already enacted and legally required to pay for, has forced us into borrowing ever more capital to keep our public services going. We borrow this money from international

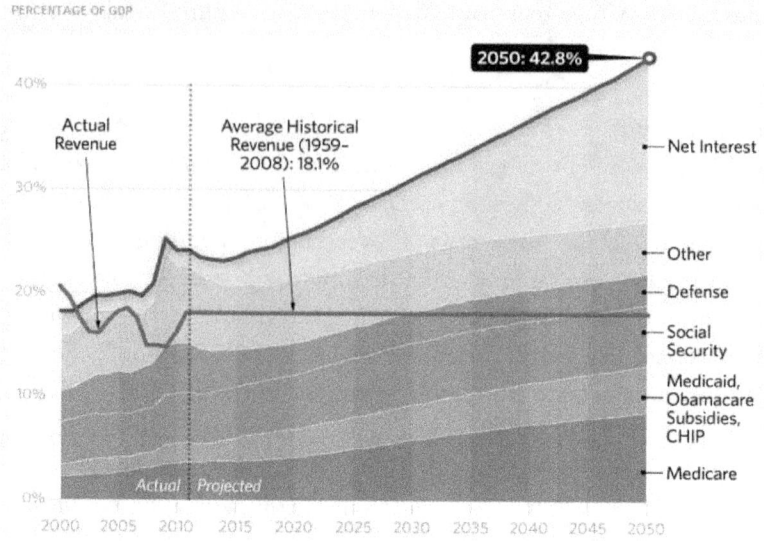

PERCENTAGE OF GDP

40%

Actual Revenue

Average Historical Revenue (1959–2008): 18.1%

2050: 42.8%

Net Interest

30%

Other

20%

Defense

Social Security

Medicaid, Obamacare Subsidies, CHIP

10%

Medicare

Actual   Projected

0%

2000 2005 2010 2015 2020 2025 2030 2035 2040 2045 2050

investors like China who see the USA as a good investment because we have among the lowest amount of debt compared to our GDP of most developed countries, and we have been politically stable for hundreds of years (the Civil War excepted). That will change as our amount of debt continues to increase in the 2020s and 2030 given our current tax policy. In the figure, the band called 'Other' includes all of our scientific research, NASA, and many other public programs.

What this means for NASA is that we are probably living, right now, in the Golden Years of NASA's budgets and in the future it may become dramatically harder to justify even the current level of investment in space activities. If Apollo is a valid historical model, it may be that once we land our first astronauts on Mars in 2035, we politically declare 'victory' and immediately start defunding these activities before we can create a permanent base on Mars. If that happens, not only will we have won, and then abandoned, our first lunar outpost by 1972, but we will have won, and then abandoned, our first interplanetary outpost on Mars in 2035.

Because of the vast distances involved, a manned expedition to Jupiter is entirely out of the question for the next political Space Spectacular. Mars may well turn out to be our first and last interplanetary venture for the remainder of this century and beyond if we turn our backs on the small incremental costs of keeping this outpost going at a time when budget and debt debates are likely to be intense. Having wrangled an asteroid…or two…and built our first Mars outpost, it is unlikely we can maintain a steady series of supply launches to Mars to keep this outpost going on anything like the current NASA budget unless we gut the entire science portion of the NASA budget. This requires a sustained political commitment, and an annual budget perhaps 20% higher than the current one, maintained through Congressional and Presidential elections every two years. Good luck with that!

These are all very hard choices to make. The next 20 years will tell whether we have reached a critical mass, economically, to continue our fledgling interplanetary adventures, or whether we finally turn our backs on this 'high frontier' having reached Mars, by declaring them too expensive to support through government programs against the rising national indebtedness. Meanwhile, you can actually donate money to NASA! When you send that donation to them, they can use it for whatever they want and you have no control over where your money goes. But, this donation will be used to fund NASA programs, not other government agencies.

There is, however, the hope that commercialization will open up access to the other 70% of the US GDP so that Mars and interplanetary activities can be funded by private companies. This is an exciting prospect because 'all' it requires is that the Consumer see a personal value to interplanetary travel that is as compelling as purchasing bottled water, cars, flat screen TVs and consumer electronics. It is worth noting that not long after NASA launched the first communications satellite in the early-1960s, the business community (who built these research satellites) immediately saw the necessity in selling trans-Atlantic TV coverage to the public. By today, the commercial satellite industry is a $200 billion annual enterprise. Some of the largest adopters of satellite communication are people living in remote areas where landlines and cell towers are not available. A simple 18-inch satellite dish and a local source of electricity (solar panels) are all that are needed for African villagers to now be connected to the world wide web.

African village connected to the internet via satellite (Image credit: BT.com)

Some Common Ideas

So what are some of the reasons that have been put forward for aggressively considering interplanetary travel?

Scientific exploration is certainly at the top of this list of reasons. Through the activities of NASA, ESA, JAXA and the Russian space agency, we have launched dozens of spacecraft beyond Earth's orbit to gather data and images about many of the larger bodies in the solar system. These efforts are all taxpayer-supported in the same way that the public supports other research efforts such as controlled fusion, alternative energy, medical research etc. Scientific exploration returns new information about our immediate environment in the universe, and has slowly made such diverse environments as the surface of Mars and the ring system of Saturn a comfortably familiar part of humanities backyard. The development of sophisticated rovers like Curiosity, complete with sophisticated imaging and laboratory facilities, is the vanguard of new generations of robotic probes that will eventually explore many of the accessible surfaces across the solar system. The search for organic, pre-biotic chemistry on Mars, Titan and the subsurface oceans of the Jovian satellites will continue in the decades to come.

61

Interplanetary Travel

Survival of humans against asteroids etc. Stephen Hawking said that he has predicted the extinction of the human race within the next thousand years unless we build habitats in space or on other planets/moons in the next two hundred years or so. That's quite a statement, and with the current economic problems facing many developed countries around the world, it is highly unlikely that any big projects such as a 10,000-person O'Neill cylinder colony will be started anytime soon. By Hawking's estimation 'We are toast'.

Economic value of resources - John Lewis in *Mining the Sky: Untold Riches from Asteroids, Comets, and Planets* estimates that the current market value of the metals in the asteroid 3554 Amun, one small nearby asteroid, is about $20 trillion. There's $8 trillion worth of iron and nickel, $6 trillion worth of cobalt, and about $6 trillion in platinum-group metals. Once we can easily launch thousands of people into orbit, and build giant solar power satellites, it shouldn't be too difficult to retrieve asteroid 3554 Amun and other asteroids like it, to supply Earth with all the metals we will ever need. The problem is, how do you recover the costs of the mining operation from the sale of the ores? Currently, these ores cost less than $1000 per gram but space-mined ores will have to cost many hundreds of thousands of dollars per gram to cover the costs of the rockets, technology, crews and other expenses launched from Earth. Consumers will not stand for the price of their goods taking a huge upward leap just because they use space-mined resources.

Space tourism – Virgin Galactic will charge $250,000 per person to take a sub-orbital flight. Already 640 passengers have signed up and paid a down payment. For some space enthusiasts, the cost is negligible. Others, though, have taken second mortgages on their homes to pay for the tickets. *SpaceShipOne* will go almost straight up 100 km to get into space, and then came nearly straight down again. Surprisingly, the first paying orbital tourists have already flown. The Russians have taken Dennis Tito and Mark Shuttleworth to the International Space Station (ISS) developed by the US, Russia, Canada, Europe, Japan and other partners. However, even at $20 million a trip, this business only makes economic sense because the international partners spent tens of billions of dollars developing the ISS for other reasons. Successful orbital mass tourism will mean not only people, but solar power satellites can be launched from the ground to orbit affordably.

Entertainment – Lockheed Martin will support the *MarsOne* mission, which is an audacious venture launched by Dutch entrepreneur Bas Lansdorp to colonize the red planet. Although Lansdorp's has indicated that Lockheed would build *Mars One's* first unmanned lander, it was later made clear that the U.S. aerospace titan had only signed up to produce a $250,000 "mission concept study" for the spacecraft, the online science journal phys.org reports. *Mars One's* estimated $6 billion project of putting humans on Mars and making them star in the first interplanetary reality TV show has gained massive attention, with 200,000 applicants willing to make the one-way journey. Currently, the mission has been pushed back two years to 2025. Meanwhile, the goal is still to put 24 people on the surface of Mars beginning in 2023. Already, tens of thousands have applied, and the winners will be trained for eight years before being deployed in groups of four starting in 2023. The people they will be sending to Mars will be ordinary people. Lansdorp explains, *"Mars One is a mission representing all humanity, and its true spirit will be justified only if people from the entire world are represented."* This is the stance that many private corporations are taking: anyone should have the opportunity to take part in this project, in the spirit of space exploration.

Helium-3 mining for fusion power –Unlike Earth, which is protected by its magnetic field, the Moon has been bombarded with large quantities of Helium-3 by the solar wind. It is thought that this isotope could provide safer nuclear energy in a fusion reactor, since it is not radioactive and would not produce dangerous waste products. The Apollo program's own geologist,

Harrison Schmidt, has repeatedly made the argument for Helium-3 mining. China has expressed strong interest in mining the moon for He3 as its ultimate long-term goal.

Solar power generation – Electrical power is a multi-hundred billion dollar per year business today. We know how to generate electricity in space using solar cells. For example, the ISS provides about 80 kilowatts continuously from an acre of solar arrays. By building much larger satellites out of hundreds of acres of solar arrays, it is possible to generate a great deal of electrical power. This can be converted to microwaves and beamed to Earth to provide electricity with absolutely no greenhouse gas emissions or toxic waste of any kind. If transportation to orbit is inexpensive following development of the tourist industry, much of Earth's power could be provided from space, simultaneously providing a large profitable business and dramatically reducing pollution on Earth. However, constructing these power satellites will be very expensive and requires far larger crews than the construction of the International Space Station.

Political reasons – Only the United States has sent humans beyond Earth's orbit, but China and India are investing in space. China became the third nation to independently launch a human into orbit in 2003 and has grown its capabilities over the last decade. The People's Liberation Army is seen as a central driver of the Chinese space program, whose ambitions include reaching the moon and building a space station by 2020. Meanwhile, India launched its first unmanned mission to Mars in late 2013, and its probe entered Mars's orbit in September 2014. The Indian Space Research Organization has since reached an agreement with NASA on subsequent explorations of Mars. "*I'm convinced that in this century the nations that lead in the world are going to be those that create new knowledge. And one of the places where you have a huge opportunity to create new knowledge will be exploration of the universe, exploration of the solar system, and the building of technology that allows you to do that,*" said former congressman and aerospace expert Robert Walker at a CFR meeting on space policy in 2013.

Religious reasons - University of Dayton political science Professor Joshua Ambrosius, used data from the General Social Survey and three Pew surveys to compare knowledge, interest and support for space exploration among Catholics, Evangelicals, Mainline Protestants, Jews, Eastern religions and those with no religion. Ambrosius found that Evangelicals — who account for one-quarter of the U.S. population — are the least knowledgeable, interested and supportive of space exploration, while Jews and members of Eastern traditions were most attentive and supportive. *"This research finds evidence that religion shapes space and space policy attitudes, even if the significant effects are dampened or eliminated once socio-demographic factors, like education, are held constant. One tradition — Evangelical Protestants — stood out as exhibiting less space knowledge, space policy support, appreciation for space exploration, and expectations of achievements in space,"* Ambrosius noted in his paper. *"All in all, this research shows that further study must be done to test the influence of religion on space exploration attitudes and, ultimately, to assess the roles of religion in our space-faring future. Ultimately, religions must ensure their survival by embracing space."*

Despite all of these plausible reasons, the federal funding of the ground-breaking technologies to make any one of them a reality remains bogged down in finding 'A Reason' that makes clear and compelling political sense. In this calculus, the words of dreamers and futurists are not welcome. When a manned expedition to Mars in the far-off 2030s is politically stalemated by the question 'Why are we REALLY doing this?' the self-evident answers drawn from any one of the possibilities above is not what is being asked. The sub-text answer must involve some dire emergency that compels the effort, like the coupling of the Atom Bomb scares of the 1950s with Soviet orbiting satellites and ICBMs. Only that seems to have the political capital to overwhelm critics at least for a short while.

However, one of the consequences of living in a largely peaceful world is that no such convenient existential answers are available. Instead, the compelling answers have to be drawn from the 2/3 of the activity that drives the US economy: Consumerism. If you can couch the need for space exploration in terms of a consumer goods or service, you can access three times the US GDP invested in government programs. This miracle has happened before with the commercial satellite industry. Perhaps the next generation of business entrepreneurs will finally figure out how to sell interplanetary space to consumers. In 2012, U.S. retail e-commerce sales amounted to $225 billion and are projected to grow to $434 billion in 2017. A survey of the most popular things that people buy online comes down to five categories:

(Image credit: Mashable.com)

Apart from clothing, all of these categories are luxuries that we have gotten by with historically, but now see as a necessity for our modern lifestyles. The challenge for space travel and exploration is to find yet another luxury that can be grown into a necessity and then tap into the incredible supply of money flowing through our economy. Companies like Space X, Bigelow and Virgin Galactic are working very hard to create new consumer-focused business models that support space travel. While hosting lucrative contracts from NASA to resupply the ISS, SpaceX is also hard at work on a trip to Mars. Bigelow has fabricated inflatable habitat modules, now about to be tested on the ISS. Meanwhile, Virgin Galactic is developing 'joy rides to near-Earth space' as a market to attract passengers at $250,000 a ticket.

In 2015, Congress passed the Space Act, which will allow private companies to start mining in space if they can find a means to do so. So it only makes sense that American lawmakers would seek to guarantee property rights for U.S. space corporations. Under the Space Act, businesses that do asteroid mining will be able to keep whatever they dig up:

Any asteroid resources obtained in outer space are the property of the entity that obtained such resources, which shall be entitled to all property rights thereto, consistent with applicable provisions of Federal law.

This is how we know commercial space exploration is serious. The opportunity here is so vast that businesses are demanding federal protections for huge, floating objects they haven't even surveyed yet. *"The bill preserves the FAA's ability to regulate commercial human spaceflight in order to protect the uninvolved public, national security, public health and safety, safety of property, and foreign policy,"* the House Science Committee said in a press release. *"It also preserves FAA's ability to regulate spaceflight participant and crew safety as a result of an accident or unplanned event."*

Meanwhile, NASA's decisive investment in the commercialization of space during the Obama Administration is starting to have its impact on space operations. Once the Space Shuttle flights were ended in 2010 and American astronauts had to pay $70 million to travel to the ISS onboard Russian launch vehicles, Congress finally understood the long term political and security implications of this arrangement and approved a massive investment in commercial launch vehicles. This approach has led to companies such as SpaceX delivering cargo to the ISS, and now at the threshold of becoming human-qualified to bring astronauts to the ISS starting in 2017. Like the beginnings of the satellite industry in the 1960s, we are going to see huge commercial pressure to offer all kinds of new services to the government and the general public as these commercial ventures begin to mature. By the time the first explorers leave for Mars, the commercial space services industry may offer trips to orbiting habitats built by Bigelow, or opportunities we can scarcely imagine. There were, for instance, no concepts like cell phones and the internet available as commercial ideas in the 1960s. The logo-rich space vehicles seen in *2001:A Space Odyssey* are finally coming to pass albeit 20-30 years later. In 2016, Bigelow's inflatable habitat will be attached to the ISS to check its long term space worthiness. Once established, Bigelow will have the green light to partner with other commercial launch services to offer a destination in space for tourists in other more lavishly designed habitats.

"The International Space Station Program will take the next step in expanding a robust commercial market in low-Earth orbit when work continues Wednesday, May 27, 2015, to prepare the orbiting laboratory for the future arrival of U.S. commercial crew and cargo vehicles. NASA Television will provide live coverage of the activity beginning at 8 a.m. EDT. NASA is in the process of reconfiguring the station to create primary and back up docking ports for U.S. commercial crew spacecraft currently in development by Boeing and SpaceX to once again transport astronauts from U.S. soil to the space station and back beginning in 2017. The primary and backup docking ports also will be reconfigured for U.S. commercial spacecraft delivering research, supplies and cargo for the crew."

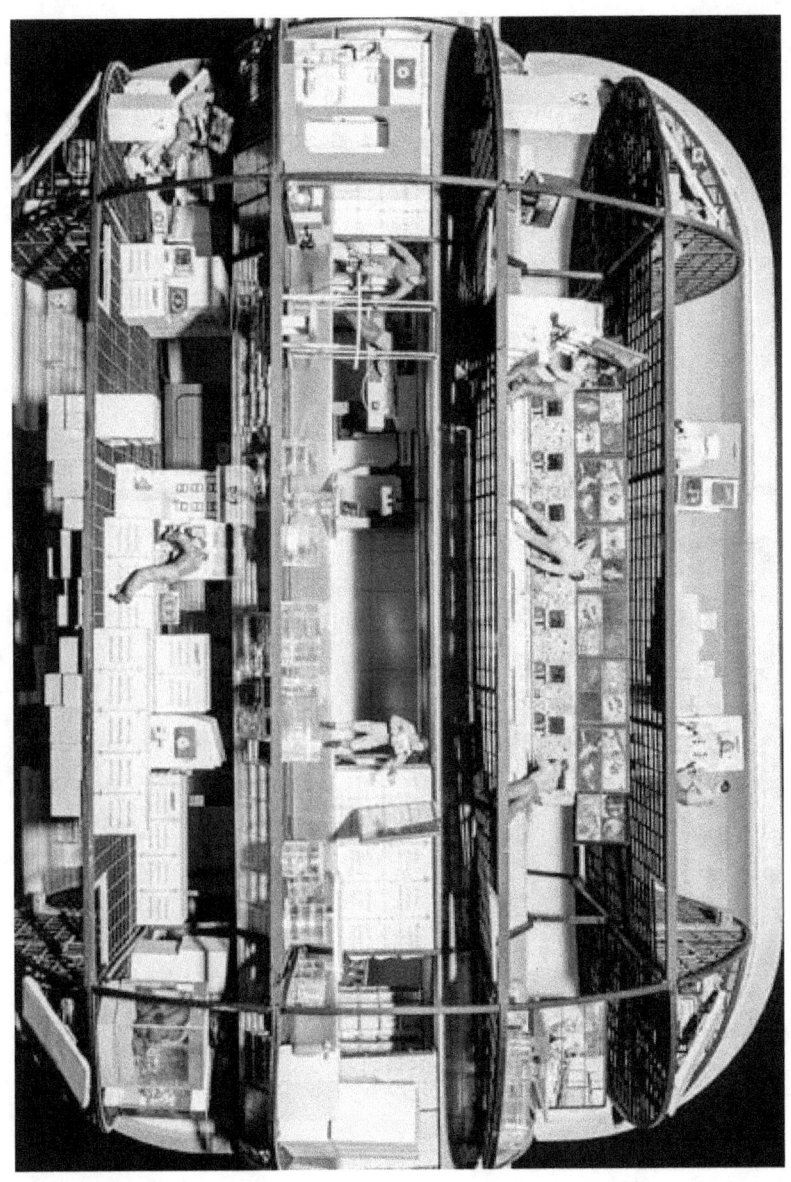

Resources and Mining

"*None of these cities in the sky was truly self-supporting, nor perhaps ever would be; but the ravenous appetite of Earth's industries for power metal and for the even more valuable planetary-core materials for such uses as jet throats and radiation shields – this insatiable demand for what the Asteroids could yield – made certain that the miners could swap what they had for what they needed...Ceres gave them heat and light and power; all of their vegetables and much of their protein came from their hydroponic tanks and yeast vats. Single-H and oxygen came from Ceres or Pallas.*" The Rolling Stones p. 206 Robert Heinlein 1952.

The first mention of asteroid mining in science fiction is apparently Garrett P. Serviss' story *Edison's Conquest of Mars*, New York Evening Journal, 1898. The 1979 film *Alien*, directed by Ridley Scott, is about the crew of the Nostromo; a commercially operated spaceship on a return trip to Earth hauling a refinery and 20 million tons of mineral ore mined from an asteroid. C. J. Cherryh's novel, *Heavy Time* focuses on the plight of asteroid miners in the Alliance-Union universe, while *Moon* is a 2009 British science fiction drama film depicting a lunar facility that mines the alternative fuel helium-3 needed to provide energy on Earth.

At the top of any list of plausible reasons is the tantalizing economic benefit of finding more resources to feed our burgeoning human population in the post-Internet age. This is not the first time we have stood on this particular beach looking out at the distant and mysterious horizon.

The main reason humans traveled and explored in past ages was to find new resources to bring back home and hopefully become wealthy. The basic idea was that the money invested in building the conveyance (ship etc), hiring a crew, and supplying the food and other resources would be handsomely paid back by whatever gold or resources could be found. Sometimes all that was needed was the promise that at the destination you could hunt, fish and build encampments to serve as a new home or basecamp for further explorations. This familiar, and expected, human understanding is completely turned on its head by the demands of space exploration and what some people call colonization.

The bottom line to any space venture is that it is hugely expensive. The most common expectation is that you will be able to arrive at a destination and find resources there that help maintain your existence there, or that can be extracted and serve to recoup the cost of the trip once sent back to Earth and sold. Although space seems like an abundant and limitless source of new resources, the reality is much more complicated. Moreover, we don't have to guess about what we will find when we get there. We already know. But first we have to master the units that geologists use for describing ore concentrations.

71

Interplanetary Travel

Image credit: Pat Rawlings/NASA/Zuma Press

We are all familiar with using percentages to indicate proportions, for example, Earth's atmosphere by volume consists of 78% nitrogen and 21% oxygen, with the remaining 1% in trace gases such as water vapor and carbon dioxide. This kind of measuring scale works well when we are talking about large fractions of a whole (100% = 1.000), but percentages become very awkward to use when we are discussing minute 'trace' constituents. Instead, we use a 'parts-per' scale. For example, the nitrogen abundance of 78% is also 78 parts per hundred. Instead of writing the abundance of carbon dioxide as 0.04% (0.0004) of the atmospheric constituents, we can write this as 4 parts per 10,000 or better yet, 400 parts per million (ppm). The conversions are also pretty simple: 0.0001% = 1 ppm = 1000 parts per billion (ppb).

For decades, astronomers have gathered meteorites and analyzed their compositions. The sample also includes rare meteorites from Mars and the moon! Over the course of many decades of study, bench-marking what they have analyzed by comparing with the rare meteorites with known orbits, and studying the reflected light from asteroidal bodies in space, a detailed classification scheme has been created that places asteroids into specific geological groups.

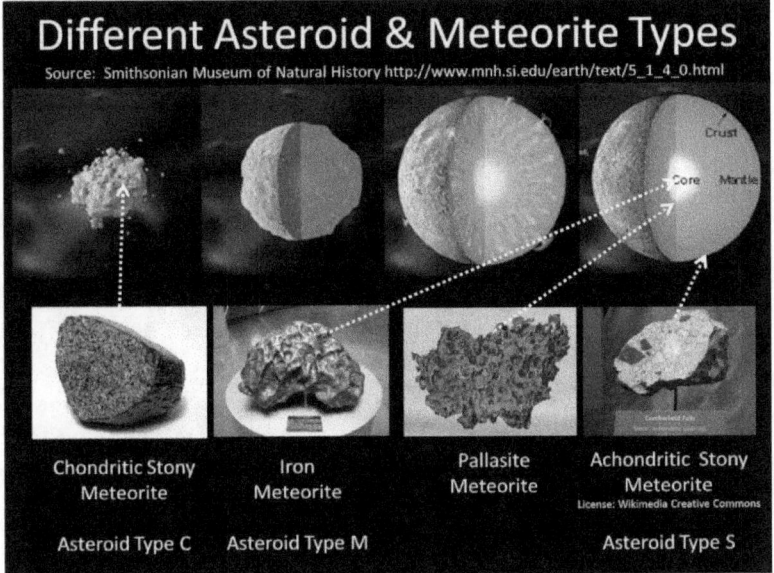

Carbonaceous 'C-type' asteroids contain lots of carbon and carbon compounds. Analysis has turned up pre-biotic chemicals, but very few heavy elements. They are very dark objects and hard to spot telescopically.

Stony 'S-types' contain lots of silicate compounds and are easy to spot as asteroids because they reflect light very well.

C-type asteroids called chondrites, are stony non-metallic meteorites that have not been modified by differentiation in the parent object. They are crumbly and composed primarily of various silicates, with an Fe-Ni free metal content between 0.3 and 35%. Chondrites are often classified according to their free metal content.

Metal or 'M-type' asteroids are rich in nearly pure iron and nickel that is industry-grade, though these asteroids and meteorites are rarer than the S and C-types.

Additional classification schemes based on the spectroscopic properties of asteroids are also in use. For example, the SMASS system is based on the spectra of 1440 asteroids and places them into 24 distinct groups based on the strength or absence of certain features in the spectra. This is analogous to the way astronomers began to classify stars into the O,B,A,F,G,K and M types early in the 20th century, primarily at Harvard Observatory.

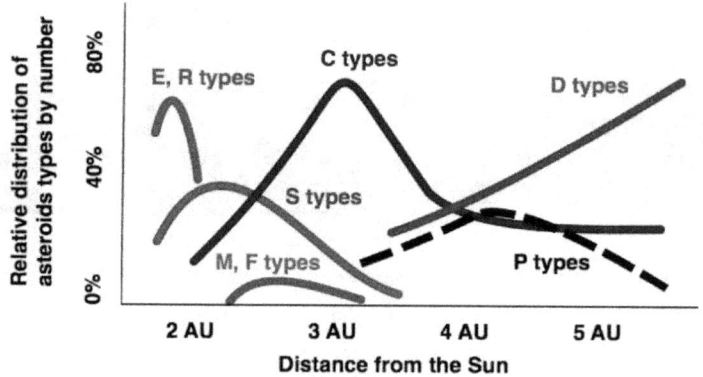

Among the C-type chondrites there are "Enstatites" (E) with around 35% free Fe-Ni granules; "High irons" (H) which average about 19% Fe-Ni; "Low-irons" (L) which average 9% Fe-Ni; and "Low iron, low metals" (LL) and "high iron, low metals" (HL) that reflect different abundances of free metal versus metal oxides, in the neighborhood of 5% Fe-Ni granules plus about 15% to 30% iron oxide in minerals (e.g., magnetite, silicates). The nonmetal ingredients of meteorites are mostly silicates, oxides and sulfur minerals: silica (35% and 40%), magnesia (20% to 25%), aluminum (2% and 3%), calcia (2%) and Iron (6%).

"C1 carbonaceous chondrites" average about 10% water in a clay mineral matrix, 2% to 5% carbon in the form of graphite, hydrocarbons and organic compounds, several percent sulfur in elemental, iron sulfide and water soluble sulfate forms, some nitrogen and other volatiles, and 5% to 15% magnetite.

"C2 carbonaceous chondrites" have very little magnetite, a little less water, carbon, and sulfur, and about 10% soluble sodium and magnesium salts, all in a mineral assemblage.

"C3, C4 and C5 carbonaceous chondrites" are very poor in water, carbon and other volatiles, but resemble C1 and C2 carbonaceous chondrites.

The reported gold contents of meteorites range from 0.0003 to 8.74 parts per million. This means in the worst cases, you have to mine about 10,000 tons of material in order to recover at least 3 grams of gold. Gold is siderophilic (iron-loving), and the greatest amounts in meteorites are in the iron phases. By comparison, the gold content of Earth's crust is in the range of 0.001 to 0.006 parts per million. That means the same 10,000 tons will get you about 60 grams of gold at the highest concentrations.

Some economic analyses usually show that the cost of returning asteroidal materials to Earth far outweighs their market value, and that asteroid mining will not attract private investment at current commodity prices and space transportation costs.

But some resources are far more valuable if they remain in space. For example, the delivery of asteroids rich in water ice to low Earth orbit could then be used for rocket fuel and save us $10,000 per pound in shipping rocket fuel from Earth's surface.

	Price/gm	Meteorite Name	
		Cape Hope	Tawallah
Cobalt	$0.04	7820 ppm	8158 ppm
Phosphorous	$0.25	445	1762
Chromium	$0.003	261	69
Cesium	$11.00	55	18
Iridium	$23.00	35	17
Platinum	$37.50	30	30
Molybdenum	$0.44	25	37
Nickel	$0.08	15	18
Palladium	$25.40	6	11
Rhenium	$16.00	5	5
Tungsten	$0.03	3.5	3.3
Rhenium	$27.50	4	1.6
Copper	$0.10	1.5	1.9
Vanadium	$2.20	0.9	0.09
Arsenic	$3.20	0.2	0.9
Gallium	$2.20	0.2	0.2
Germanium	$1.20	0.05	0.08
Gold	$39.40	0.06	0.15

Some calculations have estimated that a small metallic asteroid (1.6 km) contains more than $20 trillion worth of industrial and precious metals. Platinum is considered very rare in terrestrial geologic formations and therefore is potentially worth bringing some quantity back to Earth for terrestrial use. Platinum is valued at about $1,300 per ounce.

It should be noted that, far beyond the industrial value for asteroidal material is the collection price and novelty. Recovered meteorites from major falls typically can run from $2 to $10 per gram and even as high as $350 per gram for a Martian meteorite.

Generally, meteorite fragments run from tens to thousands of grams and are sold in one piece. These prices from collectors are generally much greater than the value of the materials in the meteorite. Samples of mined asteroids may well fetch far more than their mineral worth due to their novelty and exotic historical natures.

What would it be like to mine an asteroid? What kind of surfaces do they have? The answer to these questions will determine how much effort the mining process will have to make in just gathering the raw rock in order to crush it and extract the desired elements and compounds.

First, the masses of asteroids are so small that their gravitational pulls are very feeble. It takes a speed of 11,000 meters/sec to leave Earth, but for asteroids like Eros the escape speed is only 10 meters/sec (22 mph). Eros is a mid-sized asteroid about 34x11 km in size, with a mass of about 7 trillion tons, which is nevertheless not nearly enough to give it an appreciable gravity. In most cases, you can easily jump off the surface directly into orbit or space.

The NEAR spacecraft made a landing, or actually a docking with Eros, at a speed of 4 mph, and was able to continue sending information for two weeks before its batteries gave out. NEAR's portrait of Eros - a solid, undifferentiated, primitive relic from the solar system's formation - has already answered fundamental questions about common S-type asteroids. NEAR's 160,000 images have shown that asteroids can be incredibly diverse objects. NEAR scientists spotted more than 100,000 craters, about 1 million house-sized (or bigger) boulders, and a layer of debris resulting from a long history of impacts. Eros is not a "rubble pile" of loosely bound pieces, but rather a consolidated object. Furthermore, the chemical information gleaned from the mission is helping us to understand how asteroids like Eros are linked to meteorite samples recovered on Earth.

Eros was already known to be a complex asteroid. One side appears to have a higher pyroxene content while the opposite side displays higher olivine content. There is no air and no evidence of water on Eros. Daytime temperature is about 100 deg. C (212 deg. F), while at night it plunges to -150 deg. C (-238 deg. F). Gravity on Eros is very weak, but sufficient to hold a spacecraft in orbit. A 100-pound (45-kilogram) object on Earth would weigh about an ounce on Eros, and a rock thrown from the asteroid's surface at 22 miles/hour (10 meters/sec) would escape into space.

This image of the surface of NEAR was taken from an altitude of 250 meters and the width of the image is 36 feet (12 meters) across. The smallest objects you can see are only as big as golf balls!

Eros is classified as being an S type asteroid; one which has a composition of iron and magnesium-bearing silicates such as pyroxene and olivine mixed together with metals such as nickel and iron. NEAR Shoemaker's X-ray spectrometer detected low levels of aluminum relative to magnesium and silicon, indicating an undifferentiated composition, and that Eros may be related to the primitive ordinary chondrites.

Apart from the rocks and boulders, the surface actually seems to be powdery or sandy, which means that it would be comparatively easy to just scoop up this material and pour it into a hopper for further grinding and processing. You will not need jack hammers or power tools on S-type asteroids to do this work! M-type asteroids, which may be more profitable because of their mixtures of high-value elements may require considerably more work!

After the material is pulverized, the material can be chemically or thermally processed to result in precipitates that concentrate the desired elements. For example, a common asteroidal element is iron, and it is useful for construction of habitats. Iron is extracted by heating, and the silicate impurities become the slag, which floats to the top in the blast furnace. The resulting iron oxide is reduced to pure iron by heating it with carbon (coke). This is a problem because carbon compounds are scarce on asteroids like Eros. But partnering the Eros resources with those of a carbonaceous asteroid would do the trick!

Extracting magnesium, another common element, is more difficult. Commercial production follows two completely different methods: electrolysis of magnesium chloride or thermal reduction of magnesium oxide. Where power costs are low, electrolysis is the cheaper method—and, indeed, it accounts for approximately 75 percent of world magnesium production. The basic reaction, $2CaO + 2MgO = 2Mg + Ca_2SiO_4$ requires temperatures of about 1,800 C (3,270 F), which in space can be provided by solar concentrators.

Asteroid mining will undoubtedly result in innovating other extraction processes that reduce the mass of economically unwanted components.

Because of the low-gravity conditions, mining operations must be firmly anchored to asteroid surfaces. Most of this work can be done robotically with automated mining equipment, requiring only periodic human visits to recover the processed material. It is expected that the mined elements will be of far greater value to space operations than if they were transported back to Earth. A kilogram of iron shipped from Earth for space fabrication currently costs thousands of dollars, but mined iron may be far cheaper to deliver to a space-based construction site.

On April 24, 2012 a plan was announced by billionaire entrepreneurs to mine asteroids for their resources. The company is called *Planetary Resources*. They also plan to create a fuel depot in space by 2020 using water from asteroids, which could be broken down in space to liquid oxygen and liquid hydrogen for rocket fuel. The plan has been met with skepticism by some scientists who do not see it as cost-effective, even though platinum and gold are worth nearly $1,800 per troy ounce. An upcoming NASA mission (OSIRIS-REx) to return just 60 grams of material from an asteroid to Earth will cost about $1 billion. How will a commercial mining company amortize the expense of the technology and labor to harvest extraterrestrial ores and then sell them on Earth at a cost that is competitive with Earth-mined ores? As it turns out, returning material to Earth is as simple and inexpensive as putting it in a capsule with a heat shield and parachute. It's the launch of goods and mining equipment from Earth's surface that is the expensive part. Both of these issues have to be considered in determining the market value of extraterrestrial ores.

In September 2012, the NASA Institute for Advanced Concepts (NIAC) announced the Robotic Asteroid Prospector project, which will examine and evaluate the feasibility of asteroid mining in terms of means, methods, and systems.

Asteroid mining (Artist concept by Denise Watt/NASA)

This artwork shows an asteroid mining mission to an Earth-approaching asteroid. A NASA-sponsored study on space manufacturing held at Ames Research Center in the summer of 1977 provided much of the technical basis for the painting. "Asteroid-1" is the central long structure and the propulsion unit is the long tubular structure enveloped by stiffening yard arms and guy wires. Solar cells running the length of the propulsion system convert the sunlight into electricity which is used to power the

81

propulsion system. During the mission these solar arrays would be oriented toward the Sun to gather maximum power. In the left foreground is an asteroid mining unit, doing actual mining work. An orbital construction platform in permanent orbit provides power, supplies depot and work volume within which work proceeds.

C-type asteroids contain huge amounts of water and other organics, which can be used for drinking water, and electrolyzed into hydrogen and oxygen for fuel. Building materials made from metals will be hard to extract from these asteroids, but it is possible that ceramics can be created to take the place of metallic wall material and bulkheads. Even comet nuclei can be mined for their water-rich volatiles. A number of cometary nuclei have already been visited and imaged in detail by NASA and ESA spacecraft. The rotation rates and low gravity make these objects a challenge to dock with.

This image of the surface of the comet nucleus 67P/Churyumov–Gerasimenko was taken by the Philae lander as it approached the surface. The dust-covered boulder at upper right is about 5 meters (16.4 feet) across. The small nodules are about the size of soccer balls. (Image Credit: ESA/Rosetta.)

Preliminary results from the Philae/Mupus instrument, which deployed a hammer to the comet after Philae's landing, suggest there is a layer of dust 10-20 cm thick on the surface. Beneath that is very hard water ice, which is possibly as hard as sandstone.

"Then a daring pioneer had found that this gravel stream was actually the richest deposit of virgin ores ever discovered. The result had been Belt mining, in which individual miners built scoops on the sterns of small one-man ships. Taking these ships into the Belt, they had slowed their speed to just under that of the flowing gravel and had opened the scoop on the stern. The gravel itself poured into the holds of the little vessels." The Bell from Infinity, Robert Moore Williams, 1968 p. 25.

The Lunar Surface

We have come a long way from thinking that the moon is made from 'green cheese'. The Apollo astronauts brought back 380 kilograms of rock samples from most of the major geologic areas on the moon. Detailed assays of this material in the 1970s basically clinched the issue of composition: It is like terrestrial surface rock but with very few heavy elements.

The lunar crust extends down to a depth of 50 km, and is the layer of the moon that scientists have gathered the most information about. It is composed mostly of compounds of oxygen, silicon, magnesium, iron, calcium, and aluminum. There are also trace elements like titanium, uranium, thorium, potassium and hydrogen. The three most abundant compounds overall are ordinary silicates (47%), iron oxides (13%) and magnesium oxides (29%). Other important compounds include aluminum oxide (6%), calcium oxide (5%) and titanium oxide (0.3%). These abundances vary depending on whether you are in the highland or mare regions. The highlands have significantly more aluminum oxide (28%) and calcium oxide (16%) for example.

Here we see Lunar Prospector maps of the moon from gamma ray spectroscopy showing where thorium deposits are found. The diamonds are the Apollo and Luna landing sites.

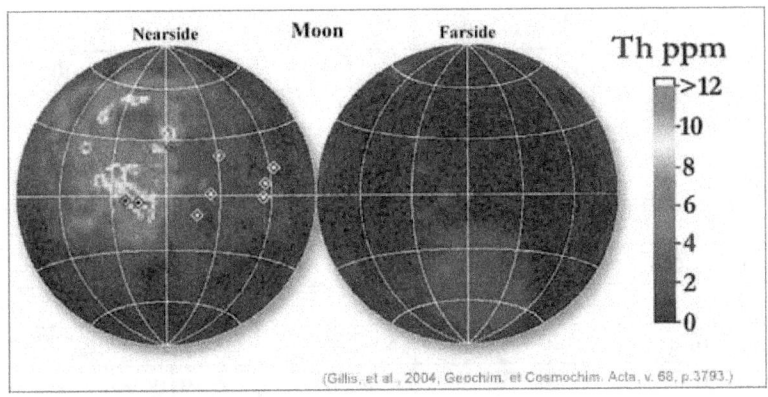

(Gillis, et al., 2004, Geochim. et Cosmochim. Acta, v. 68, p.3793.)

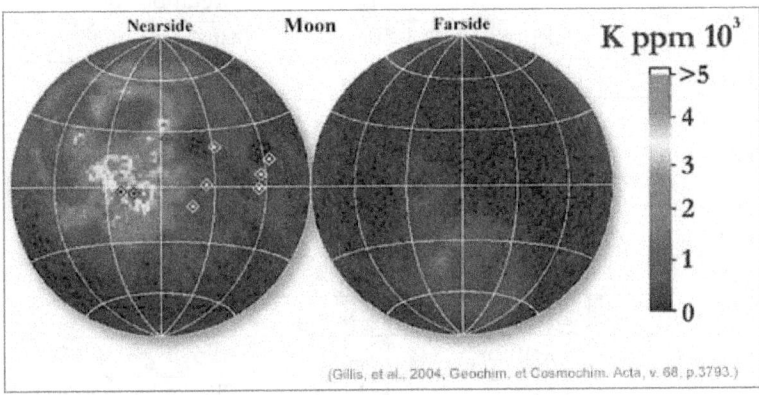

(Gillis, et al., 2004, Geochim. et Cosmochim. Acta, v. 68, p.3793.)

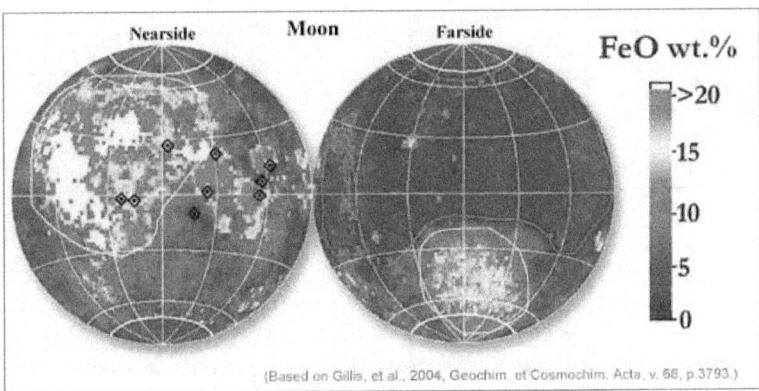

(Based on Gillis, et al., 2004, Geochim. et Cosmochim. Acta, v. 68, p.3793.)

85

Element	Abundance	Price/gm	Use
Silicon	21.9 %	$0.50	Integrated circuits
Magnesium	17.5 %	$0.50	Aluminum alloys
Iron	10.6 %	$0.0045	Building
Calcium	3.3 %	$30.00	Bio element
Aluminum	3.2 %	$0.15	Construction
Manganese	0.12 %	$0.0023	Metal alloys
Sodium	0.06 %	$0.07	Bio element
Chromium	4200 ppm	$0.07	Electroplating
Titanium	1800 ppm	$0.003	Construction
Nickel	400 ppm	$0.08	Construction
Vanadium	150 ppm	$1.38	Metal bonding
Potassium	83 ppm	$0.65	Bio element
Strontium	34 ppm	$1.00	Magnets
Ruthenium	30 ppm	$14.00	Solar cells
Palladium	22 ppm	$58.33	Catalytic converters
Cobalt	20 ppm	$0.44	Batteries, magnets
Scandium	19 ppm	$14.00	Lighting, alloys
Copper	15 ppm	$0.80	Electricity
Zirconium	17 ppm	$1.57	Anti-corrosion
Tellurium	14 ppm	$0.24	Alloying
Cesium	12 ppm	$11.00	Photoelectric cells
Barium	11 ppm	$0.55	Medicine
Rhodium	10 ppm	$25.00	Alloys, catalysis
Yttrium	6 ppm	$2.20	Red color TVs
Zinc	5 ppm	$0.27	Alloys, batteries
Cerium	3 ppm	$0.13	Alloys, catalysts
Neodymium	2 ppm	$4.20	Magnets
Gallium	1 ppm	$2.20	Electronics
Niobium	1 ppm	$0.18	Alloys
Molybdenum	1 ppm	$0.44	Steel alloying
Lanthanum	1 ppm	$0.80	Batteries, lighting
Dysprosium	1 ppm	$0.45	Neutron shielding

Element	Abundance	Price/gm	Use
Gadolinium	1 ppm	$0.45	Medicine, imaging
Lithium	0.8 ppm	$0.27	Medicine
Erbium	0.8 ppm	$5.40	Neutron absorber
Silver	0.8 ppm	$1.20	Money! Solder
Samarium	0.7 ppm	$3.60	IR absorber
Ytterbium	0.7 ppm	$14.00	Radiation source
Antimony	0.6 ppm	$0.05	Electronics
Hafnium	0.5 ppm	$1.20	Neutron absorber
Praseodymium	0.4 ppm	$4.50	Alloying agent
Europium	0.3 ppm	$1350	Neutron absorber
Holmium	0.3 ppm	$8.60	Magnets
Rubidium	0.3 ppm	$12.00	Ion propellant
Terbium	0.2 ppm	$50.40	Electronics
Beryllium	0.2 ppm	$7.48	Alloying agent
Boron	0.1 ppm	$11.14	Window glass
Germanium	0.1 ppm	$3.60	Electronics
Indium	0.1 ppm	$9.68	Electronics
Thulium	0.1 ppm	$70.00	Radiation source
Lutetium	0.1 ppm	$340.00	Catalyst

The table of element abundances shows that the accessible lunar surface is very rich in important elements such as titanium and chromium in easily recoverable amounts. Also among the moon's vast riches are iron, vanadium, palladium, platinum, and tungsten. However, the elements most needed for organic chemistry (food etc) are nearly completely absent. There are no sources of carbon and nitrogen, however oxygen (60%), calcium (3%) and iron (11%) seem to be in good supply.

One hundred parts per million is equivalent to 0.01% of a substance. At concentrations of 1 part per million, you will have to process 1 ton of rock to recover one gram of the element. If we were to use the previous table and fully extract all of the minerals we found in one ton of rock, the recovered minerals would amount to $1.2 million, with calcium ($990,000) and silicon ($109,000) accounting for the most valuable components. Of course, the relevant cost is shipping these minerals from Earth at a cost of $10,000 per pound, which works out to an additional $10 million added to the $1.2 million per ton of minerals. It is cheaper to mine a pound of calcium out of lunar rocks than to have it shipped from Earth.

NASA has recently teamed up with heavy-equipment giant Caterpillar Inc. to develop drilling and mining technologies that could one day be used in space. Of particular interest is efficient ways to extract key materials like water ice and a variety of precious metals embedded in large asteroids, which could be worth millions back here on Earth. Metals like iron, nickel, titanium, platinum group metals are crucial to fuel-cell and clean-tech projects. Rare-earth elements such as lithium are used in batteries and mobile phones. The moon also has the isotope helium-3, a fuel source for future nuclear fusion plants.

For many years, NASA has also offered a mining competition for university-level students. Now in its sixth year, students are challenged to design and build a mining robot that can traverse the simulated Martian chaotic terrain, excavate Martian regolith and deposit the regolith into a collector bin within 10 minutes.

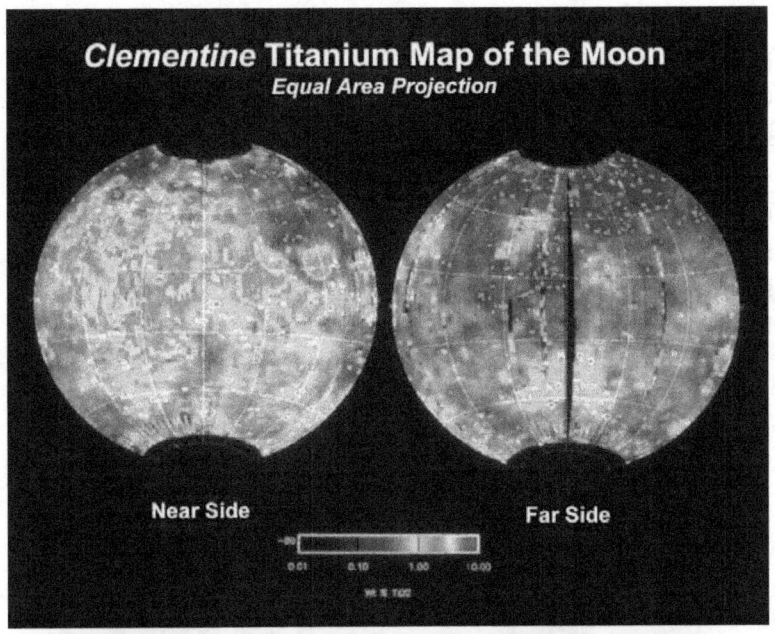

This image of the moon is from NASA's Moon Mineralogy Mapper on the Indian Space Research Organization's Chandrayaan-1 mission. It is a three-color composite of reflected near-infrared radiation from the sun, and illustrates the extent to which different materials are mapped across the side of the moon that faces Earth. Small amounts of water and hydroxyl in the polar regions were detected on the surface of the moon at various locations. (Image Credit: ISRO/NASA/JPL-Caltech/Brown Univ./USGS)

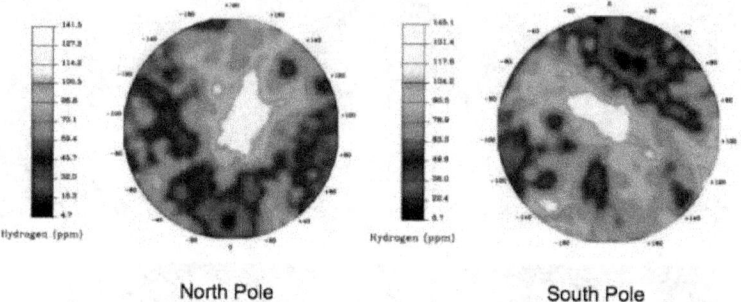

North Pole   South Pole

The Lunar Prospector used its neutron spectrometer maps to create maps of the hydrogen emission at the lunar poles. These low resolution data, however, do not tell us the form of the hydrogen. The most plausible form is water ice based upon radar reflections of these regions. (Image credit: D. Lawrence, Los Alamos National Laboratory)

The presence of ice on the moon is a game changer. If true, it would be the mining focus of a colony providing much needed water for human survival, and fuel for rockets.

Using data from a NASA radar that flew aboard India's Chandrayaan-1 spacecraft in 2010, ice deposits have been detected near the moon's north pole. The synthetic aperture radar found more than 40 small craters with water ice. The craters range in size from 1 to 9 miles (2 to15 km) in diameter. Although the total amount of ice depends on its thickness in each crater, it's estimated there could be at least 1.3 trillion pounds (600 million metric tons) of water ice.

In 2009, the Lunar Crater Observation and Sensing Satellite (LCROSS) was a robotic spacecraft attached to the Lunar Reconnaissance Orbiter. The main LCROSS mission objective was to explore the presence of water ice in a permanently shadowed crater near a lunar polar region. It was successful in discovering water in the southern lunar crater Cabeus by impacting with the wall of the crater and creating a gas cloud (see picture), whose spectrum indicated an abundance of water equal to 6% by mass of the regolith. (Image credit: NASA/LCROSS)

Currently, the lunar South Polar Region with its confirmed deposits of water ice are considered the prime landing spots for future rovers and prospectors, and perhaps human visitations. With a significant deposit of water ice, astronauts have a local source of water and can also generate oxygen for breathing and liquid hydrogen and oxygen for rocket fuels.

NASA researchers have already field tested a hydrogen reduction plant and lunar rover prospectors on Hawaii's volcanic soil. During a year-long operation, the reduction plant produced 1,455 pounds (660 kg) of oxygen from a rocky soil containing 5 percent iron oxide. "*You can make back costs fairly quickly compared to the launch costs of just throwing tanks of water and oxygen at the moon*," said Gerald Sanders, manager of NASA's In-Situ Resource Utilization Project.

Here we see the latest prototype of a real Moon drill called DESTIN, which stands for Drilling Exploration & Sample Technology INtegrated, that may be chosen to spearhead NASA's lunar prospecting mission in 2020 if the mission makes it to launch. The agency has spent about $20 million on the project by 2014, but expects its investment to top out around a quarter of a billion dollars. (Image credit: NASA)

NASA plans to launch Resource Prospector in 2020. After a 3-day journey from Earth to the moon, the lander will set down on the lunar surface in the polar regions and deploy a rover. The rover will use prospecting tools to search for sub-surface water, hydrogen and other volatiles. When an appropriate location is found, a drill will extract samples of the lunar regolith from as deep as one meter below the surface. The sample will be heated in an oven to determine the type and quantity of elements and compounds such as hydrogen, nitrogen, helium, methane, ammonia, hydrogen sulfide, carbon monoxide, carbon dioxide, sulfur dioxide – and most importantly, water!

Resource Prospector for lunar mining (Image credit: NASA)

Another lunar mining resource is helium-3, which is a light, non-radioactive isotope of helium with two protons and one neutron. The abundance of helium-3 is thought to be greater on the Moon (embedded in the upper layer of regolith by the solar wind over billions of years), though still low in quantity from one to 50 ppb is helium-3. Current US industrial consumption of helium-3 is approximately 60,000 liters (approximately 8 kg) per year. The cost at auction has typically been approximately $100/liter although increasing demand has raised prices to as much as $2,000/liter in recent years. Materials on the Moon's surface contain helium-3 at concentrations on the order of between 1.4 and 15 ppb in sunlit areas, and may contain concentrations as much as 50 ppb in permanently shadowed regions. Obtaining helium-3 from lunar regolith will not be an easy task. You would have to refine millions of tons of lunar soil before gathering enough helium-3 to be useful in fusion reactions on Earth.

Apollo 16 astronaut stands near the rim of Plum Crater, 30-meters in diameter. If you replace the lunar rover with a deluxe moon truck, and the astronauts walking stick with a shovel or mining tool, you get some idea of what the lunar surface mining experience might be like. (Image credit: NASA).

Mercury Resources

In Arthur C. Clarke's *Rendezvous with Rama* (1973), Mercury is ruled by a hot-tempered government of metal miners that threatens to destroy the alien spacecraft Rama. Ben Bova's *Mercury* (2005) is about the human drama of the exploration of Mercury: why people might be interested in going there (for instance, to harness the intense solar energy that close to the sun), and what challenges there would be. In Stephen Baxter's *Manifold: Space* (2000), Mercury is the final stronghold of humanity after successive waves of extraterrestrial colonizers exterminate the human race from  the rest of the solar system. It is the site of the human victory against a race of beings known as the Crackers, who seek to dismantle Sol to exploit its energy. Finally, Mark Anson's 2011 novel *Below Mercury* is a science fiction thriller set in the abandoned

95

workings of Erebus Mine, an ice mining and refining facility in the depths of the Chao Meng-fu crater, on the South Pole of Mercury.

The first images of Mercury's surface by the Mariner 10 flybys in 1974 revealed it to be a darker version of our moon with a variety of cratered land forms, as well as some unique geologic formations. Astronomers have estimated that the Mercury composition is made up of approximately 70% metals and 30% silicate material. Apparently this massive core contracted as it cooled and caused the entire surface to buckle like a sundried raisin.

The MESSENGER spacecraft was launched in 2004 and after a number of gravity assists, finally settled into orbit around Mercury in 2011, and spent the next 4 years mapping the planet at high resolution and performing composition studies using a gamma ray and x-ray spectrometer. What was noted almost immediately, and what you can see in the four colorized images, was that unlike the significant color and mineralogical differences seen on the lunar surface from the highland and mare rocks, there were very few large contrasts to be found on the surface of Mercury. This suggested that the iron-rich compounds found in the surface of the moon were largely absent from Mercury. The lack of detectable iron oxides and silicates implies that Mercury's surface consists mostly of iron-poor, calcium-magnesium silicates. MESSENGER's neutron spectrometer showed that the surface abundance of iron plus titanium is comparable to that of several lunar mare regions.

Iron - The most recent values for the iron content of Mercury's surface suggest that the average surface has about 1.5 percent iron oxide (FeO). Most lunar highland rocks contain about 6 % or more FeO, and the lunar maria can have over 20 % FeO. The current favored model of Mercury's interior is that it has a solid iron sulfide (FeS) layer up to 200 km thick that sits on top of the iron core. The FeS layer would lie beneath Mercury's outer silicate mantle and crust, which together are several hundred kilometers thick, and would never be exposed by typical impact craters or basins.

Magnesium - The distribution of magnesium varies wildly from place to place. In some places it is as common as silicon, but in "fresh" exposed areas near young craters there seems to be very little. Mercury's surface seems to be dominated by high-Mg minerals such as enstatite, plagioclase feldspar, and lesser amounts of Ca, Mg, and/or Fe sulfides, which explains the difference between the volcanic smooth plains and the older surface areas of Mercury. We think that the differences have to do with the crystallization of the smooth plains from a more chemically evolved magma source over 3 billion years ago. Mercury's surface

97

is Mg-rich but Al- and Ca-poor compared with typical terrestrial and lunar crustal material. The older terrains on Mercury have higher ratios of magnesium to silicon, sulfur to silicon and calcium to silicon than the northern plains do, but it also has lower ratios of aluminum to silicon.

In this image we see a data strip from the MESSENGER's X-ray spectrometer that shows how the abundance of magnesium falls off dramatically in a slightly bluish deposit ejected from a recent crater.

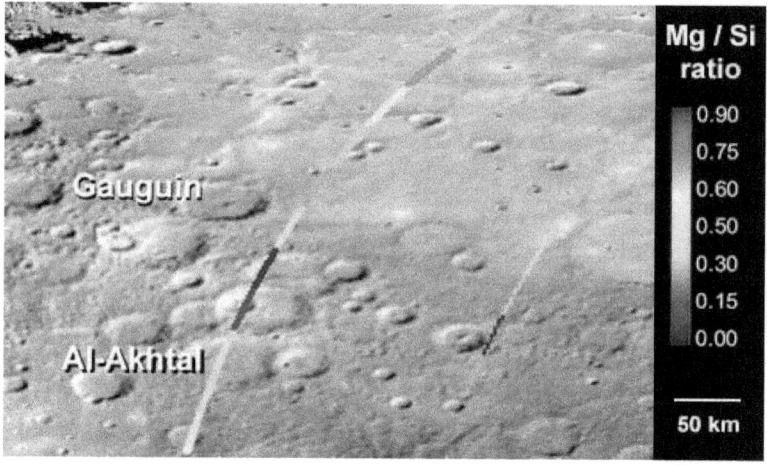

Sodium - The MESSENGER data also show that up to about 5% of Mercury's soil is rich in the element sodium and is bound in the structures of minerals that make up the rocks such as plagioclase feldspar ($NaAlSi_3O_8$), commonly found in many igneous rocks. The most abundant sulfide minerals on Mercury's surface are probably oldhamite (CaS) and niningerite (MgS).

Water - MESSENGER's Neutron Spectrometer data are consistent with hydrogen (in the form of water ice) within shadowed craters at Mercury's north and south poles. These were originally discovered in the 1990s with radar studies of polar surface reflectivity, which was interpreted to indicate subsurface ices were present. MESSENGER scientists estimate that the polar ice amounts to about 20 billion to 1 trillion tons.

Calcium - The majority of the calcium on Mercury's surface is probably bound into minerals like plagioclase feldspar. It is possible that at least some of the calcium on Mercury's surface may also be found in one or more non-aluminous, non-sulfide minerals such as diopside ($MgCaSi_2O_6$).

Element	Northern Planes	Cratered Terrain
Magnesium	8.5% by weight	14.4% by weight
Iron	4%	4%
Calcium	3.7%	4.9%
Sulfur	1.5%	2.3%
Aluminum	6.6%	5.4%
Potassium	2000 ppm	500 ppm

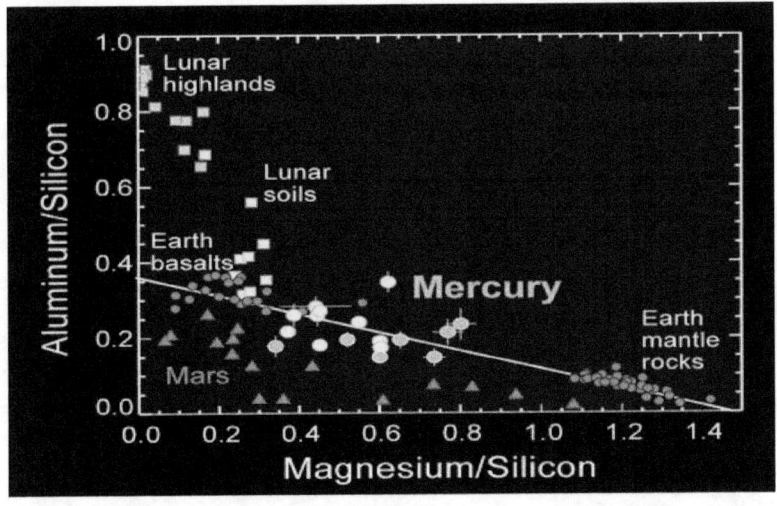

Potassium, uranium - Mercury has a higher abundance of potassium than expected. So far there is no indication of large amounts of uranium on Mercury's surface. Potassium is a volatile element and thorium is a refractory one, so the ratio of these two elements indicates how much heating the material has been exposed to. For example, the ratio for the Moon (360) is much lower than that for Earth (3000), because of the intense heating during the formation of the moon. The ratio for Mercury (~6000), is similar to the other planets which means that the formation of Mercury was an event similar to that of Earth, Venus and Mars and did not involve high temperatures.

MESSENGER has also provided information on the concentrations of potassium, thorium, uranium, sodium, chlorine, and silicon. There is a large terrain on Mercury covering some 5 million square kilometers that exhibits the highest observed magnesium/silicon, sulfur/silicon, and calcium/silicon ratios, as well as some of the lowest aluminum/silicon ratios on the planet's

100

surface, and could be the site of an ancient impact basin. Maps of magnesium/silicon (left) and thermal neutron absorption (right) across Mercury's surface (red indicates high values, blue low). These maps, together with maps of other elemental abundances, reveal the presence of distinct geochemical terrains. Volcanic smooth plains deposits are outlined in white.

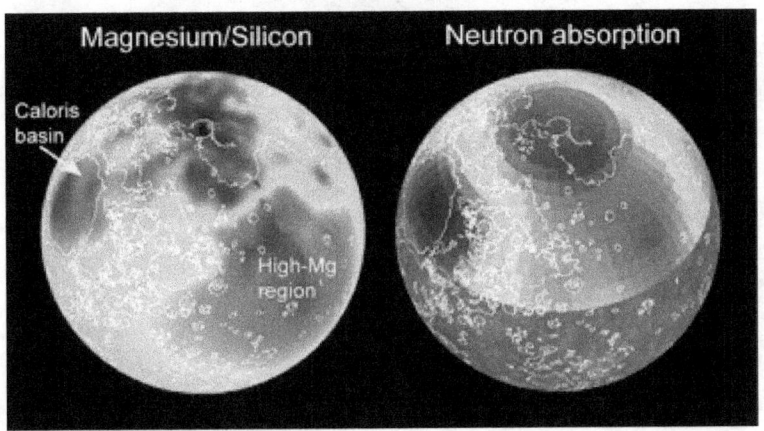

So far as we can understand today, Mercury is relatively barren of many interesting industrial or biological minerals, but it does have a sizable deposit of water ice, which is a valuable commodity in the inner solar system. Although we can't write off Mercury as a site for future mining activities, it seems unlikely that anyone would want to go to all the trouble to get there, only to extract minerals that can be found on our own moon at far less cost!

At an incredible 5 meters per pixel, this is one of the highest-resolution images of Mercury's surface ever captured. It was acquired on March 15, 2014 with the MESSENGER spacecraft's MDIS (Mercury Dual Imaging System) instrument and shows an 8.3-km (5.2-mile) -wide section of Mercury's north polar region, speckled with small craters and softly rolling hills. The smallest crater is about 20 feet across; big enough for a small house with a dome!

On April 30, 2015, the MESSENGER spacecraft sent its final image. The image shown here is the last one acquired and transmitted back to Earth by the mission. Captured by the mission's Narrow Angle Camera (NAC) of the Mercury Dual Imaging System (MDIS), this view spans only about 1 mile wide, with a resolution of 2.1 meters per pixel. In the view appears to be a relatively flat terrain dotted with small craters and smooth lumps. The image is located within the floor of the 93-kilometer-diameter crater Jokai. The spacecraft struck the planet just north of Shakespeare basin. (Image Credit: NASA/Johns Hopkins University Applied Physics Laboratory/Carnegie Institution of Washington)

Mining Mars

It would be easy to write an entire book about the science fiction stories that feature Mars and Mars mining. It remains one of the most popular solar system locations to begin just about any science fiction story since the start of the 20th century.

Mars surface by Spirit Rover near McMurdo. (Credit NASA)

Little was known about the surface of Mars before the arrival of the Mariner 4 spacecraft in 1965, which snapped a few dozen pictures of its cratered terrain. For me as a 12-year-old in 1964, they were very disappointing pictures that showed no canals or alien waterways like so many science fiction stories had promised. During the next 50 years, many spacecraft have meticulously mapped its surface and used a variety of instruments to determine the composition of its rocks and minerals.

Silicon - Silicon is one of the most abundant elements on the surface of both Mars and Earth (second only to oxygen). The most extensive region of highest silicon content is located in the high latitudes north of Tharsis (centered near 45 degrees latitude, -120 degrees longitude). The area of lowest silicon content lies just to the east of the Hellas Basin (-45 degrees latitude, 90 degrees longitude).

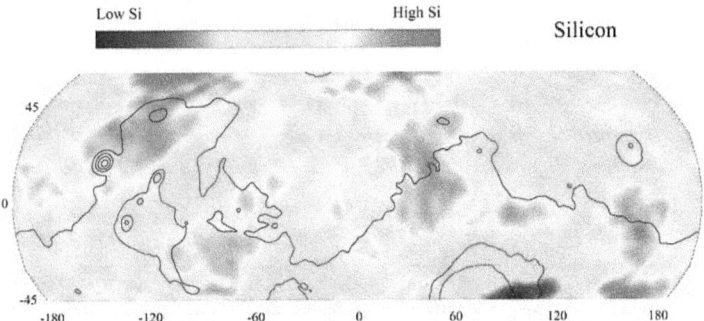

Mars is fundamentally an igneous planet. Rocks on the surface and in the crust consist predominantly of minerals that crystallize from magma including feldspar, pyroxenes and olivine, and similar to the "weathered basaltic soils" of Hawaiian volcanoes. The dark areas of Mars are abundant in the minerals olivine, pyroxene, and plagioclase feldspar. These minerals are the primary constituents of basalt. The mineral olivine occurs all over the planet, but some of the largest concentrations are in Nili Fossae, an area containing Noachian-aged rocks. Another large olivine-rich outcrop is in Ganges Chasma, an eastern side chasm of Valles Marineris.

Pyroxene minerals are also widespread across the surface. Both low-calcium and high-calcium pyroxenes are present, with the high-calcium varieties associated with younger volcanic shields and the low-calcium forms (enstatite) more common in the old highland terrain.

On Earth, vast volcanic landscapes are called Large Igneous Provinces such as the Tharsis region, are sources of nickel, copper, titanium, iron, platinum, palladium, and chromium. It has for some time been accepted by the scientific community that a group of meteorites recovered from the surface of Earth actually came from Mars. In these meteorites, called SNCs, many important elements have been detected, and magnesium, aluminum, titanium, iron, and chromium are relatively common in them. In addition, lithium, cobalt, nickel, copper, zinc, niobium, molybdenum, lanthanum, europium, tungsten, and gold have been found in trace amounts. It is quite possible that in some places these materials may be concentrated enough to be inexpensively mined.

This image shows the Viking 2 landing area. The rounded rock in the center foreground is about 20 centimeters wide. The angular rock to the right and further back than the rounded rock is about 1.5 meters across. There are two trenches that were dug in the regolith to the right of the rounded rock, as well as one behind and slightly to the left. The gently sloping troughs between the artificial trenches and the angular rock, which cut from the middle left to the lower right corner of the picture, are natural surface features. (Credit NASA/Viking)

The Mars landers Viking I, Viking II, Pathfinder, Opportunity Rover, and Spirit Rover identified aluminium, iron, magnesium, and titanium in the Martian soil. Opportunity found small structures, named "blueberries" that were found to be rich in hematite, a major ore of iron. These blueberries could easy be gathered up and reduced to metallic iron that could be used to make steel. In addition, the Spirit, Opportunity and Curiosity

Rovers have found nickel-iron meteorites sitting on the surface of Mars. These could also be used to produce steel. (Image credit: NASA/Curiosity)

In December 2011, Opportunity Rover discovered a vein of gypsum sticking out of the soil. Tests confirmed that it contained calcium, sulfur, and water. The vein, called "Homestake," is in Mars' Meridiani plain. Homestake is in a zone where the sulfate-rich sedimentary bedrock of the plains meets older, volcanic bedrock exposed at the rim of Endeavour crater. Dark sand dunes are common on the surface of Mars. Their dark tone is due to the volcanic rock basalt. The basalt dunes are believed to contain the minerals chromite, magnetite, and ilmenite. Since the wind has gathered them together, they do not even have to be mined, merely scooped up. These minerals could supply future colonists with chromium, iron, and titanium.

We now think that the most abundant chemical elements in the Martian crust, besides silicon and oxygen, are iron, magnesium, aluminum, calcium, and potassium. These are perfect for creating the raw materials for human habitats so that we don't have to import them from Earth at huge cost. All we need is to establish a manufacturing base, that could be robotically controlled to create or even '3-D print' whatever colonists need.

WATER MAP
2001 Mars Odyssey Gamma Ray Spectrometer

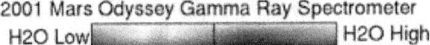

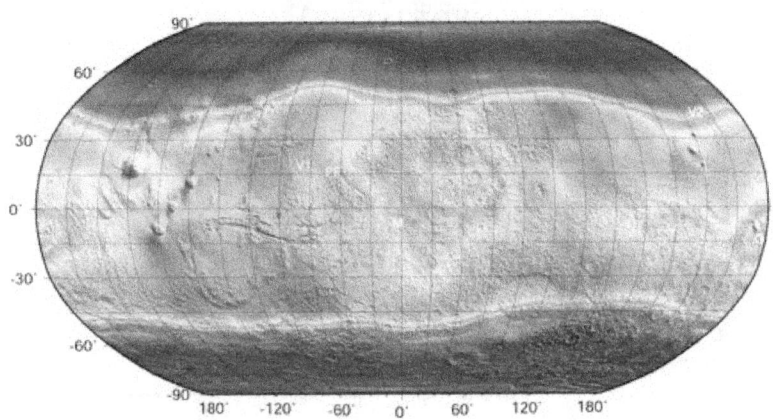

Potassium Map
2001 Mars Odyssey Gamma Ray Spectrometer

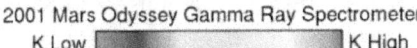

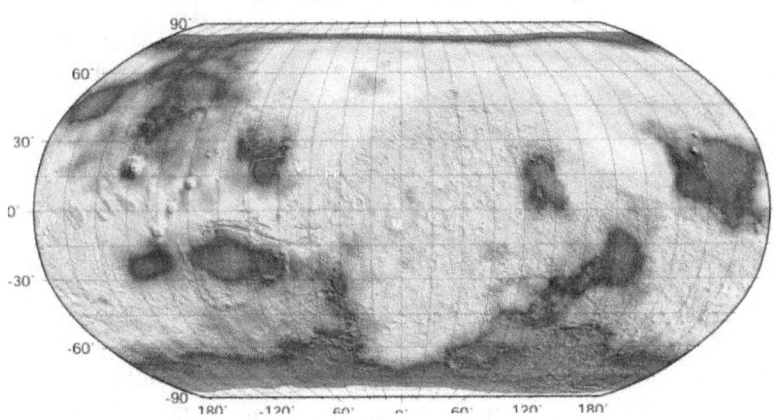

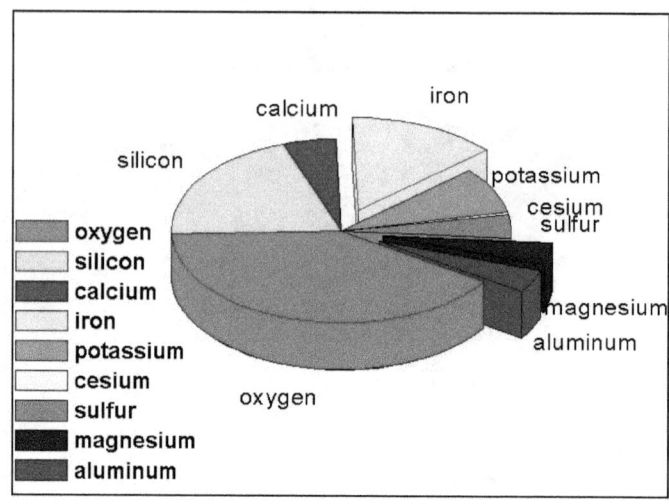

The elements titanium, chromium, manganese, sulfur, phosphorus, sodium, and chlorine are less abundant but are still important components of many accessory minerals in rocks and of secondary minerals in the dust and soils (the regolith). Hydrogen is present as water (H_2O) ice and in hydrated minerals. Carbon occurs as carbon dioxide (CO_2) in the atmosphere and sometimes as dry ice at the poles. An unknown amount of carbon is also stored in carbonates. Molecular nitrogen (N_2) makes up 2.7 percent of the atmosphere. As far as we know, organic compounds are absent except for a trace of methane detected in the atmosphere.

This gamma ray spectrometer map of the mid-latitude region of Mars is based on gamma-rays from the element thorium. Thorium is a naturally radioactive element that exists in rocks and soils in extremely small amounts. The region of highest thorium content, shown in red, is found in the northern part of Acidalia Planitia (50 degrees latitude, -30 degrees longitude). Areas of low thorium content, shown in blue, are spread widely across the planet with significant low abundances located to the north of Olympus Mons

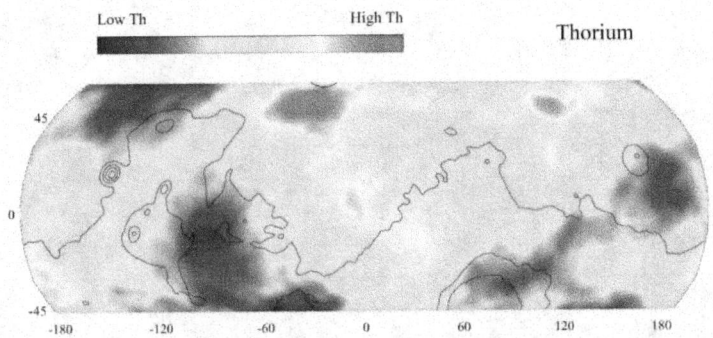

(near 55 degrees latitude, -155 degrees longitude), to the east of the Tharsis volcanoes (-10 degrees latitude, -80 degrees longitude) and to the south and east of Elysium Mons (20 degrees latitude, 160 degrees longitude). Contours of constant surface elevation are also shown. The long continuous contour line running from east to west marks the approximate separation of the younger lowlands in the north from the older highlands in the south.

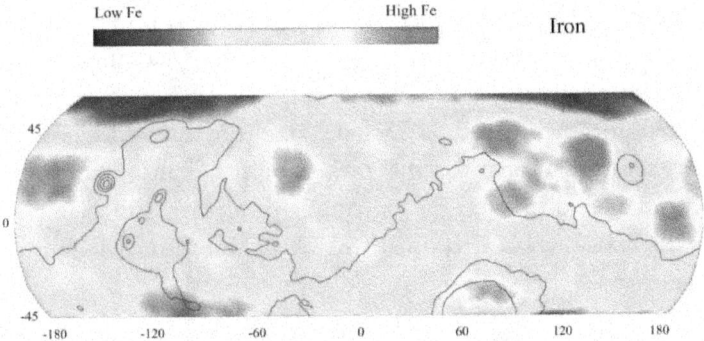

This gamma ray spectrometer map of the mid-latitude region of Mars is based on gamma-rays from the element iron. Iron is among of the most abundant elements on the surface of both Mars and Earth, and is responsible for the red color on the surface of Mars. Regions of highest iron content are concentrated in the area spanning from Utopia Planitia to Amazonis Planitia (right and left sides of the map) and within Acidalia Planitia (just left of center).

111

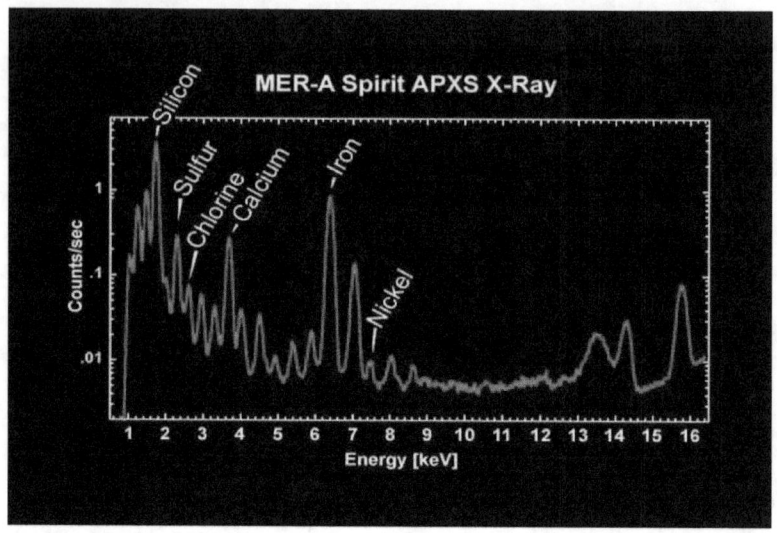

This graph or spectrum taken by the alpha particle X-ray spectrometer onboard the Mars Exploration Rover Spirit shows the variety of elements present in the soil at the rover's landing site. Iron and silicon make up the majority of the Martian soil. Sulfur and chlorine were also observed as expected. Trace elements detected for the first time include zinc and nickel. These latter observations demonstrate the power of the alpha particle X-ray spectrometer to pick up the signatures of elements too faint to be seen before. (Image Credit: NASA/JPL/Max-Planck-Institute for Chemistry)

The next graph shows the abundances of elements in the Martian rock "Jake Matijevic" as detected by the Curiosity rover. The Jake rock is low in magnesium and iron, high in elements like sodium, aluminum, silicon and potassium, which often are in feldspar minerals. It has very low nickel and zinc.

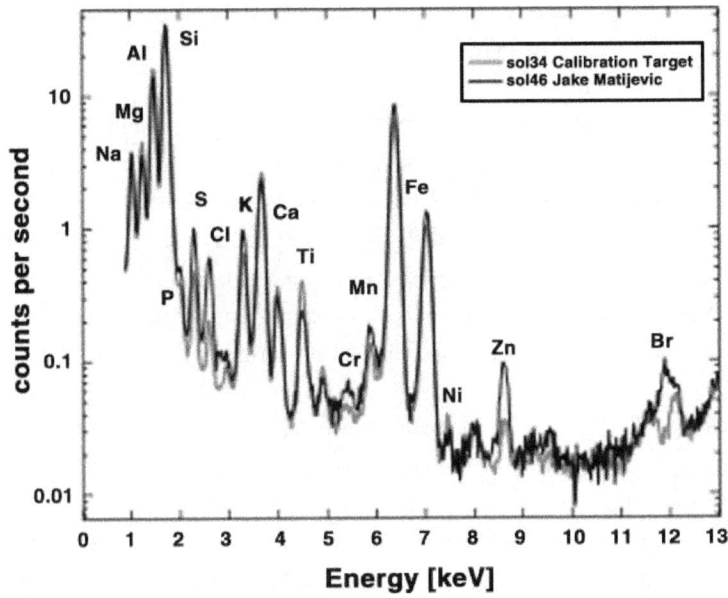

The salt-forming elements sulfur, chlorine and bromine are likely in soil or dust grains visible on the surface of the rock. These results point to an igneous or volcanic origin for this rock. (Image credit: NASA/JPL-Caltech/University of Guelph/CSA)

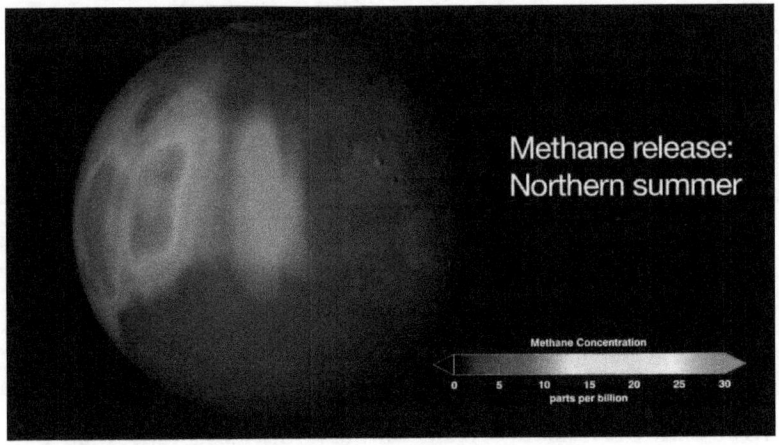

Methane release:
Northern summer

Methane Concentration

0   5   10   15   20   25   30
parts per billion

Interplanetary Travel

Methane - The debate about methane on Mars started way back in 1969 when scientists working with the Mariner 7 spacecraft announced that the spacecraft had likely detected methane with its Infrared Spectrometer Instrument. The announcement created a huge stir because of its implications to life on Mars, however, project scientists publicly retracted the finding several weeks later after completing additional analysis of the spectra acquired by Mariner 7 that revealed that the unexpected band at a wavelength of 3.3μm that was mistaken for methane was actually created by carbon dioxide ice in an unexpected configuration.

Then in 2009, satellite measurements revealed several plumes of methane release during the northern mid-summer totaling 19,000 tons at a 'hot spot' concentration of about 60 ppb. Atmospheric measurements by the Curiosity rover in Guisev Crater later confirmed methane in the local atmosphere at a level of 0.7 ppb over 20 months of observations. More tantalizing was a sudden and temporary rise in methane concentration to over 7 ppb suggesting that a localized, temporary release of methane took place, indicating that Mars is active, from a geologic sense – a finding that is of great interest to scientists. Methane is definitely on Mars and has a seasonal release cycle, but what is causing it remains a mystery…and no one wants to say it is from living systems because there is as yet no on-the-ground proof of a biological origin.

Nitrogen – This element has been detected in the Martian atmosphere at a level of 1.9%, but the Curiosity rover also measured nitrates in the Martian soil, which is significant for visitors who want to grow food from Martian soil. Nitrates were also found in wind-blown dust so this compound is widespread across Mars. The drill samples show a nitrate abundance about 1,100 parts per million, detected through measuring nitric oxide.

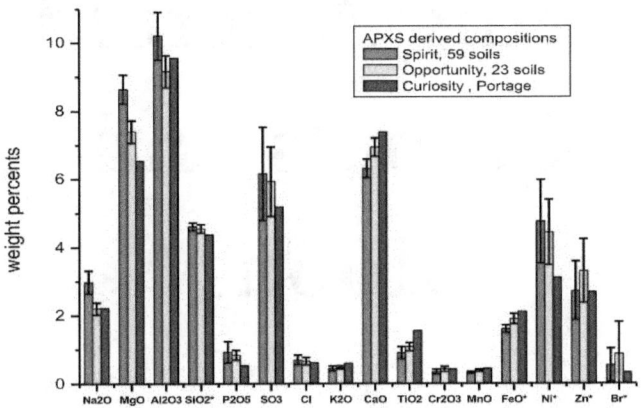

This graph compares the elemental composition of typical soils at three landing regions on Mars: Gusev Crater, where NASA's Mars Exploration Rover Spirit traveled; Meridiani Planum, where Mars Exploration Rover Opportunity still roams; and now Gale Crater, where NASA's newest Curiosity rover is currently investigating. The data from the Mars Exploration Rovers are from several batches of soil, while the Curiosity data are from soil taken inside a wheel scuff mark called "Portage" and examined with its Alpha Particle X-ray Spectrometer (APXS). (Image Credit: NASA/JPL-Caltech/University of Guelph)

Mining Asteroids, Comets and Moons

The Moons of Mars: Phobos

The two small asteroid-like bodies orbiting Mars, Phobos and Deimos, exhibit similar visible to near-infrared spectra. Phobos exhibits two distinct types of materials across its surface, and data from both Mars Express and Mars Reconnaissance Orbiter have provided additional details about the properties of these materials and their spatial relation to one another. An extensive regolith is observed to have developed on both moons with characteristics that may be unique due to their special environment in Mars orbit. The detailed composition of the moons of Mars remains uncertain.

The High Resolution Imaging Science Experiment (HiRISE) camera on NASA's Mars Reconnaissance Orbiter took two images of the larger of Mars' two moons, Phobos, within 10 minutes of each other on March 23, 2008. This is the second, taken from a distance of about 5,800 kilometers (about 3,600 miles). In the full-resolution version of this image, a pixel encompasses 5.8 meters (19 feet), providing a resolution (smallest visible feature) of about

116

15 meters (about 50 feet). It is presented in color by combining data from the camera's blue-green, red, and near-infrared channels.

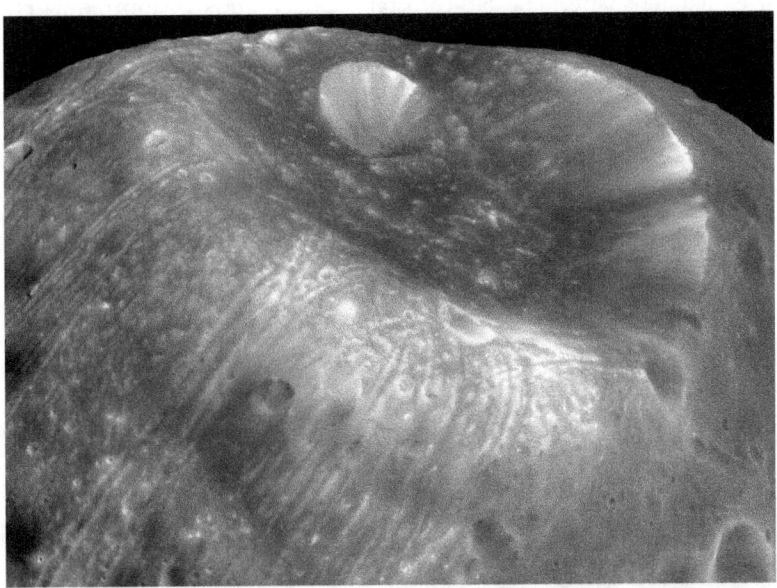

The most prominent feature in the images is the large crater Stickney. With a diameter of 9 kilometers (5.6 miles), it is the largest feature on Phobos. Materials near the rim of Stickney appear bluer than the rest of Phobos. This could mean this surface is fresher, and therefore younger, than other parts of Phobos. Troughs and crater chains are obvious on other parts of the moon. Although many appear radial to Stickney in this image, recent studies from the European Space Agency's Mars Express orbiter indicate that they are not related to Stickney. Instead, they may have formed when material ejected from impacts on Mars later collided with Phobos. The lineated textures on the walls of Stickney and other large craters are landslides formed from materials falling into the crater interiors in the weak Phobos gravity (less than one one-thousandth of the gravity on Earth).

The Moons of Mars: Deimos

Deimos is the smaller of the two Martian moons and is less irregular in shape. The largest crater on Deimos is approximately 2.3 km in diameter, 1/5 the size of the largest crater on Phobos. Although both moons are heavily cratered, Deimos has a smoother appearance caused by the partial filling of some of its craters. When impacted, dust and debris will leave the surface of the moon because it doesn't have enough gravitational pull to retain the ejecta. However, the gravity from Mars will keep a ring of this debris around the planet in approximately the same region that the moon orbits. As the moon revolves, the debris is redeposited as a dusty layer on its surface.

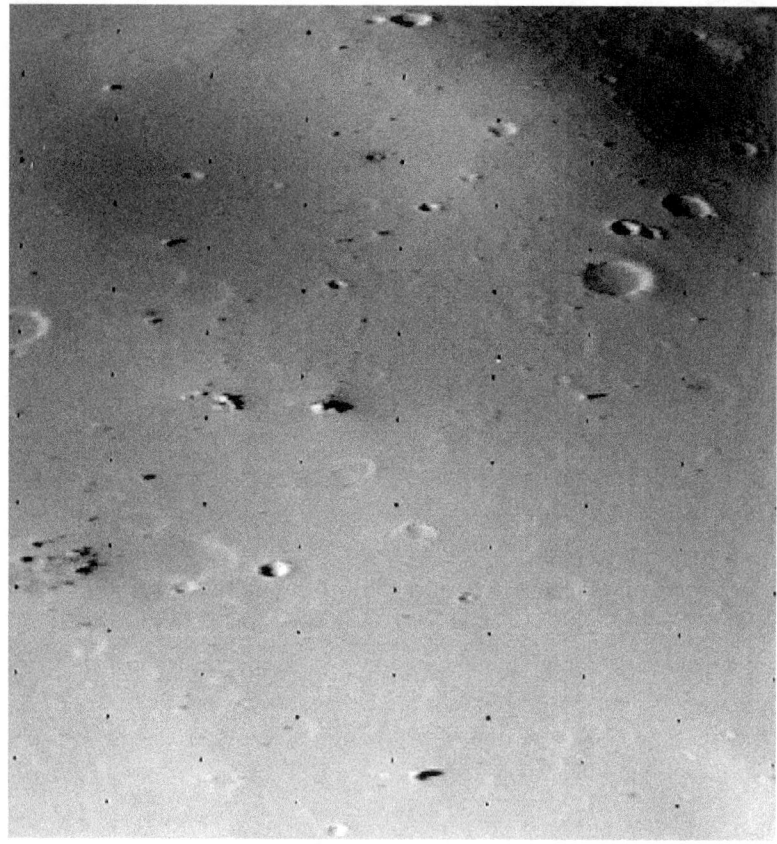

This Viking 2 image shows the surface of Deimos from a distance of 30 km. Features as small as 3 meters across can be seen. Note many of the craters are covered over by a layer of dust estimated to be about 50 meters thick. Large blocks, 10 to 30 meters across, are also visible.

Asteroid Itokawa

25143 Itokawa is an Apollo asteroid and Mars orbit-crosser. It was the first asteroid to be the target of a sample return mission by the Japanese space probe Hayabusa, and the smallest asteroid photographed by a spacecraft. Itokawa is an S-type asteroid. Radar imaging by Goldstone in 2001 observed an ellipsoid 630 m long and 250 m wide. Surface gravity 0.1 m/sec and escape speed 0.2 m/s. Mass = 35 billion kg. The Hayabusa mission confirmed these findings and also suggested that Itokawa may be a contact binary formed by two or more smaller asteroids that have gravitated toward each other and stuck together.

The Hayabusa images show a surprising lack of impact craters and a very rough surface studded with boulders, described by the mission team as a 'rubble pile'. The 26 August 2011, issue of Science devoted six articles to findings based on dust that Hayabusa had collected from Itokawa. Scientists' analysis suggested that Itokawa was probably made up from interior fragments of a larger asteroid that broke apart. Dust collected from the asteroid surface was believed to have been exposed there for about eight million years.

Scientists used varied techniques of chemistry and mineralogy to analyze the dust from Itokawa. Itokawa's composition was found to match the common type of meteorites known as "low-total-iron, low metal ordinary chondrites". Another team of scientists determined that the dark iron color on the surface of Itokawa was the result of abrasion by micrometeoroids and high-speed particles from the Sun which had converted the normally whitish iron oxide into a darker coloring. (Image credit: JAXA/Hayabusha)

This high resolution image shows evidence for material alignments that may indicate some type of surface flow mechanism. The comet's surface appearance is consistent with a regolith of rubble. The coarse debris ranges from pebble size to boulders tens of meters in diameter. This rubble is probably rearranged by occasional meteor impacts that send shock waves through the comet nucleus and shake the surface violently. This may cause the materials to sort themselves out by size.

Asteroid Vesta

Vesta is thought to consist of a metallic iron–nickel core about 220 km in diameter, an overlying rocky olivine mantle, with a surface crust. The asteroid Vesta is covered with a surprising amount of hydrogen, and bits of Vesta may have rained down on Earth in the form of meteorites.

Vesta's composition appears strikingly similar to a class of meteorites on Earth called howardite, eucrite and diogenite (HED) leading scientists to think these chunks of meteoritic rock originally came from Vesta.

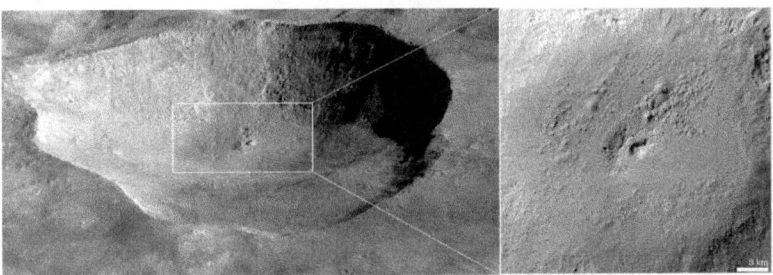

High resolution of a portion of Vesta showing details 100-meters across. (Credit: NASA/Dawn)

The formation of the 460-km-diameter impact crater on Vesta would have ejected a million cubic kilometers of material, which is dramatically more than the volume of the over 200 small asteroids that comprise the Vesta family. There is a large collection of potential samples from Vesta accessible to scientists, in the form of over 1200 HED meteorites (Vestan achondrites), giving insight into Vesta's geologic history and structure. NASA Infrared Telescope Facility (NASA IRTF) studies of asteroid (237442) 1999 TA10 suggest that it originated from deeper within Vesta than the HED meteorites.

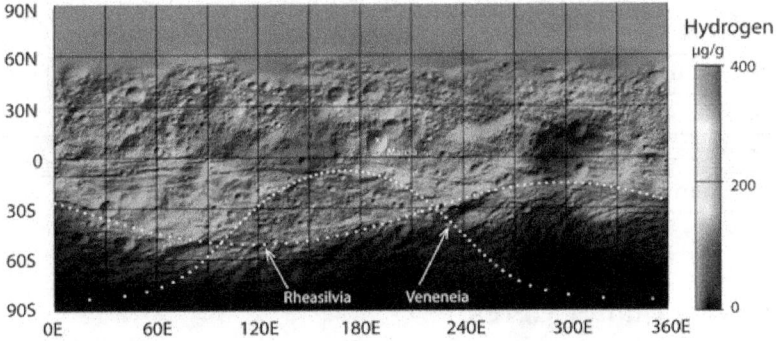

The Rheasilvia region is richest in diogenite. The presence of olivine within the Rheasilvia region would also be consistent with excavation of mantle material. Gullies found in the Marcia and

Cornelia craters may have formed by the transient flow of liquid water after buried deposits of ice were melted by the heat of the impacts. Dark surface material is probably composed of carbonaceous chondrite material deposited on the surface by impacts.

Howardites are calcium-rich achondrites. Their interiors are always display a wealth of different granules cemented in a pulverized or impact-melted matrix. Because of this, the chemical composition of Vesta deduced from the various meteoritic granules may be only a rough indicator of the mineral and element abundances in the crust of Vesta.

Material	Abundance	Material	Abundance
MgO	17.6 %	Gold	3 ppm
FeO	17.3 %	Barium	2 ppm
Al_2O_3	5.2 %	Cerium	1.6 ppm
CaO	3.9 %	Neodymium	1.1 ppm
Cr_2O_3	0.78 %	Lanthanum	0.5 ppm
TiO_2	<0.8%	Gadolinium	0.5 ppm
Na_2O	0.12 %	Yttrium	0.5 ppm
K2O	0.017%	Samarium	0.3 ppm
Vanadium	117 ppm	Hafnium	0.3 ppm
Scandium	19 ppm	Rubidium	0.1 ppm
Cobalt	12 ppm	Europium	0.1 ppm
Strontium	11 ppm	Holmium	0.1 ppm
Cesium	3 ppm	Thulium	0.1 ppm

Dwarf Planet Ceres

Ceres is the largest object in the asteroid belt. Composed of rock and ice, it is 950 kilometers (590 miles) in diameter and has a mass of 9.4×10^{20} kg. It comprises approximately one third of the mass of the asteroid belt. It is the only dwarf planet in the inner Solar System and the only object in the asteroid belt known to be unambiguously rounded by its own gravity. Ceres appears to be differentiated into a rocky core and icy mantle, and may harbor a remnant internal ocean of liquid water under the layer of ice. The surface is probably a mixture of water ice and various hydrated minerals such as carbonates and clay. In January 2014, emissions of water vapor were detected from several regions of Ceres. This was somewhat unexpected, because large bodies in the asteroid belt do not typically emit vapor, a hallmark of comets.

The surface composition of Ceres is broadly similar to that of C-type asteroids, but some differences do exist. The ubiquitous features in Ceres's IR spectrum are those of hydrated materials, which indicate the presence of significant amounts of water in its interior. Other possible surface constituents include iron-rich clay minerals (cronstedtite) and carbonate minerals (dolomite and siderite), which are common minerals in carbonaceous chondrite meteorites. The spectral features of carbonates and clay minerals are usually absent in the spectra of other C-type asteroids. Sometimes Ceres is classified as a G-type asteroid.

The Cererian surface is relatively warm. The maximum temperature with the Sun overhead was estimated from measurements to be 235 K (about −38 °C, −36 °F) on 5 May 1991. Ice is unstable at this temperature. Material left behind by the sublimation of surface ice could explain the dark surface of Ceres compared to the icy moons of the outer Solar System.

These new photos show additional differentiation between the glowing spots, and actually make the case for ice somewhat stronger. The new surface images from the Dawn spacecraft were taken from an altitude of 4,500 miles (7,200 km) with a resolution of 2,250 feet (700 meters) per pixel. For reference, the crater containing the glowing objects is about 57 miles across.

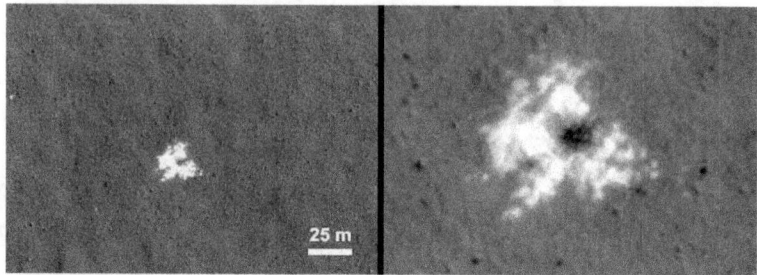

There are indications that Ceres may have a tenuous water vapor atmosphere outgassing from water ice on the surface.

Surface water ice is unstable at distances less than 5 AU from the Sun, so it is expected to sublime if it is exposed directly to solar radiation. Water ice can migrate from the deep layers of Ceres to the surface, but escapes in a very short time. As a result, it is difficult to detect water vaporization. Water escaping from polar regions of Ceres was possibly observed in the early 1990s but this has not been unambiguously demonstrated. It may be possible to detect escaping water from the surroundings of a fresh impact crater, or from cracks in the subsurface layers of Ceres. Ultraviolet observations by the IUE spacecraft detected statistically significant amounts of hydroxide ions near Ceres' north pole, which is a product of water vapor dissociation by ultraviolet solar radiation.

In early 2014, using data from the Herschel Space Observatory, it was discovered that there are several localized (not more than 60 km in diameter) mid-latitude sources of water vapor on Ceres, which each give off about 3 kg of water per second. Two potential source regions, designated Piazzi (123°E, 21°N) and Region A (231°E, 23°N), have been visualized in the near infrared as dark areas (Region A also has a bright center) by the W. M. Keck Observatory. Possible mechanisms for the vapor release are sublimation from about 0.6 km² of exposed surface ice, or cryovolcanic eruptions resulting from radiogenic internal heat or from pressurization of a subsurface ocean due to growth of an overlying layer of ice. Surface sublimation would be expected to decline as Ceres recedes from the Sun in its eccentric orbit, whereas internally powered emissions should not be affected by orbital position. The limited data available are more consistent with cometary-style sublimation.

Jupiter's Moon Io

With over 400 active volcanoes, Io is the most geologically active object in the Solar System. This extreme geologic activity is the result of friction generated by tidal heating generated within Io's interior as it is pulled between Jupiter and the other Galilean satellites: Europa, Ganymede and Callisto. Several volcanoes produce plumes of sulfur and sulfur dioxide that climb as high as 500 km (300 mi) above the surface. Io's surface is also dotted with more than 100 mountains. Some of these peaks are taller than Mount Everest.

Unlike most satellites in the outer solar system that are mostly composed of water ice, Io is primarily composed of silicate rock surrounding a molten iron or iron sulfide core. Most of Io's surface is composed of extensive plains coated with sulfur and sulfur dioxide frost. The magnetosphere of Jupiter sweeps up gases and dust from Io's thin atmosphere at a rate of 1 ton per second. This material is mostly composed of ionized and atomic sulfur, oxygen and chlorine; atomic sodium and potassium; molecular sulfur dioxide and sulfur; and sodium chloride dust.

Io's colorful appearance is the result of various materials produced by its extensive volcanism. These materials include silicates (such as orthopyroxene), sulfur, and sulfur dioxide. Sulfur dioxide frost is ubiquitous across the surface of Io, forming large regions covered in white or grey materials. Sulfur is also seen in many places across Io, forming yellow to yellow-green regions. Sulfur deposited in the

mid-latitude and polar regions is often radiation damaged, which produces Io's red-brown polar regions. Explosive volcanism, often taking the form of umbrella-shaped plumes, paints the surface with sulfurous and silicate materials. Plume deposits on Io are often colored red or white depending on the amount of sulfur and sulfur dioxide in the plume. These red deposits near the volcano called Pele consist primarily of sulfur, sulfur dioxide, and perhaps Cl_2SO_2.

Io's high density suggests that it contains little or no water, though small pockets of water ice or hydrated minerals have been tentatively identified, most notably on the northwest flank of the mountain Gish Bar Mons. Io has the least amount of water of any known body in the Solar System.

Io has an extremely thin atmosphere consisting mainly of sulfur dioxide, with minor constituents including sulfur monoxide, sodium chloride, and atomic sulfur and oxygen. The thin Ionian atmosphere also means any future landing probes sent to investigate Io will require retro-thrusters for a soft landing. The thin atmosphere also necessitates a rugged lander capable of enduring the strong Jovian radiation environment, which a thicker atmosphere would have attenuated.

Jupiter's Moon Europa

Pictures of Jupiter's moon Europa taken by the Galileo spacecraft suggest there is now, or was in the past, an ocean beneath the satellite's frozen crust. The darker areas are most likely composed of deposits of salty minerals such as sulfates and carbonates.

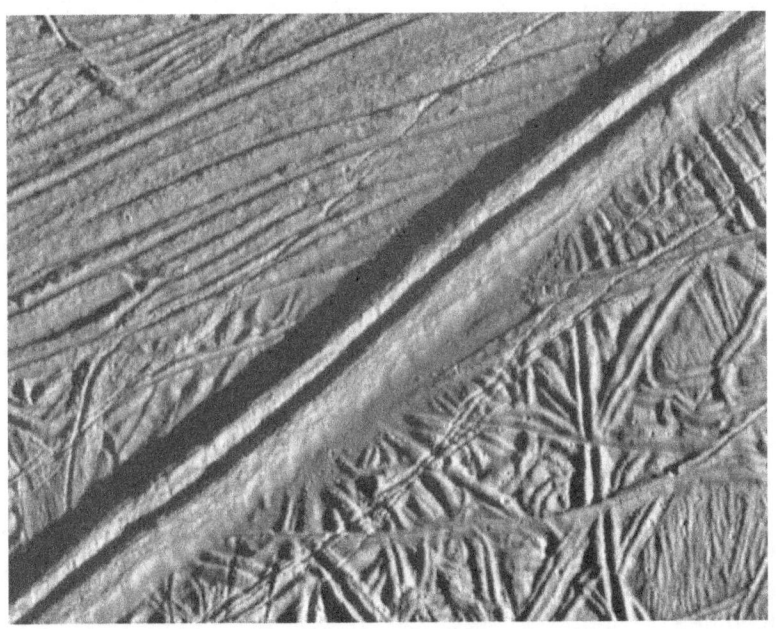

The area photographed by Galileo is about 14 kilometers by 17 kilometers (8.7 miles by 10.6 miles), and has a resolution of 20 meters (22 yards) per pixel. Illumination is from the right (east). One of the youngest features seen in this area is the double ridge cutting across the picture from the lower left to the upper right. This ridge is about 2.6 kilometers (1.6 miles) wide and stands some 300 meters (330 yards) high. Small craters are most easily seen in the smooth deposits along the south margin of the prominent

127

double ridge, and in the rugged ridged terrain farther south. (Image credit: NASA/JPL)

The best matches to the composition of the ices near the ridges were hydrated salts, including sulfates, carbonates, and borates. The best bets seem to be natron (Na2CO3(10H2O), epsomite (MgSO4(7H2O), and hexahydrite (MgSO4(6H2O), although there are a few small differences between them and the Europa spectra.

To be useful for explorers, we need to have deposits of interesting materials that are exposed to the surface so that a minimum amount of expensive and hazardous mining needs to be attempted.

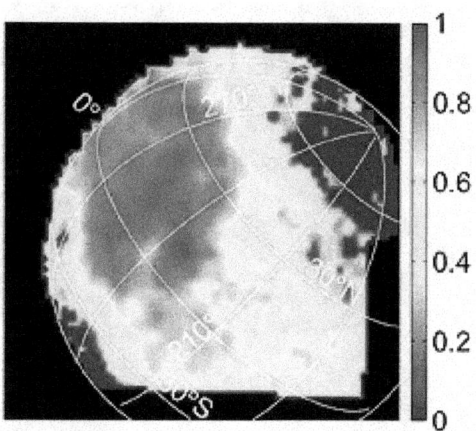

This map shows the distribution of almost pure ice (dark blue), pure non-ice (red), and mixtures of the two (colors in between) on the surface of Europa. It seems as though the surface may have deposits of materials that are not simply a form of ordinary water ice, and perhaps among this material there may be enough economically-interesting minerals to be found.

Jupiter's Moon Ganymede

Ganymede is composed of about equal amounts of silicate rock and water ice. It has an iron-rich, liquid core, and probably an internal ocean that may contain more water than all of Earth's oceans together. Its surface is composed of two main types of terrain. Dark regions, saturated with impact craters and dated to four billion years ago, cover about a third of the satellite. Lighter regions, crosscut by extensive grooves and ridges and only slightly less ancient, cover the remainder.

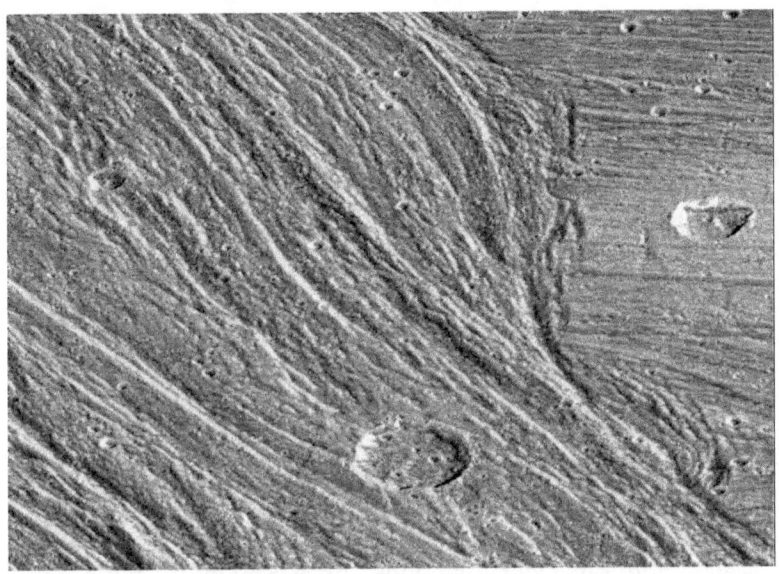

This image from the Galileo spacecraft shows complex sets of ridges and grooves are visible in this image of the Nippur Sulcus region on Jupiter's largest moon Ganymede. The image covers an area approximately 79 kilometers (50 miles) by 57 kilometers (36 miles) across. The resolution is 93 meters (330 feet) per picture element. The Nippur Sulcus region is an example of Bright Terrain on Ganymede which is typified by multiple sets of ridges and grooves. In this image, a younger sinuous northwest-southeast trending groove set cuts through, and apparently destroys, the older east-west trending features on the right of the image, allowing scientists to determine the sequence of events that led to the region's formation. The area contains many impact craters. The large crater in the bottom of the image is about 12 kilometers (8 miles) in diameter.

Ganymede's magnetosphere was probably created through convection within its liquid iron core. The satellite has a thin oxygen atmosphere that includes O, O_2, and possibly O_3 (ozone). Water ice seems to be ubiquitous on the surface, with a mass fraction of 50–90%, significantly more than in Ganymede as a whole. Galileo spacecraft data also shows signs of carbon dioxide, sulfur dioxide and, possibly, cyanogen, hydrogen sulfate and various organic compounds. Galileo results have also shown magnesium sulfate ($MgSO_4$) and, possibly, sodium sulfate (Na_2SO_4) on Ganymede's surface. These salts may originate from the subsurface ocean.

Jupiter's Moon Callisto

Callisto is composed of approximately equal amounts of rock and ices, with a mean density of about 1.83 g/cm^3, the lowest density and surface gravity of Jupiter's major moons. Compounds detected spectroscopically on the surface include water ice, carbon dioxide, silicates, and organic compounds. Investigation by the Galileo suggests a small silicate core and possibly a subsurface ocean of liquid water at depths greater than 100 km. The exact composition of Callisto's rock component is not known, but is probably close to the composition of L/LL type ordinary chondrites, which are characterized by less total iron, less metallic iron and more iron oxide than H chondrites.

Water ice seems to be ubiquitous on the surface of Callisto, with a mass fraction of 25–50%. Galileo spacecraft data reveals magnesium- and iron-bearing hydrated silicates, carbon dioxide, sulfur dioxide, and possibly ammonia and various organic compounds. Small, bright patches of pure water ice are intermixed with patches of a rock–ice mixture and extended dark areas made of a non-ice material.

Callisto's surface at 3-meter per pixel resolution (Credit NASA/JPL/Galileo)

Callisto's surface is also asymmetric: the leading hemisphere is darker than the trailing one. The trailing hemisphere appears to be enriched in carbon dioxide, whereas the leading hemisphere has more sulfur dioxide. Overall, the chemical composition of the surface, especially in the dark areas, may be close to that seen on D-type asteroids whose surfaces are made of carbonaceous material.

Saturn's Moon Iapetus

For decades, astronomers have known that one side of this moon is significantly brighter than the other. We now know that the dark material is composed of metallic iron, nano-size iron oxide (hematite), CO_2, H_2O ice, and possible signatures of ammonia, bound water, H_2 or OH-bearing minerals, trace organics, and as yet unidentified materials. The dark material has a large component of fine, sub-0.5-μm diameter particles consistent with hematite and iron.

Image of Iapetus showing remarkable surface contrasts (Image credit: NASA/ESA/Cassini)

Spectral signatures of ice also indicate that sub-0.5-μm diameter particles are present in the icy regions. The dark material on Iapetus matches those seen on Phoebe, Hyperion, Dione, Epimetheus, Saturn's rings Cassini Division, and the F-ring implying the material has a common composition throughout the Saturn system.

The ultraviolet image indicates water ice abundance across the surface: the bright north polar terrain (shown in red) is the iciest region in this view. The darkest terrain, which includes very little water ice, is shown in light blue. (Image credit: NASA/JPL)

Saturn's Moon Titan

Titan is the only moon in our solar system with a dense atmosphere. At its surface, the pressure is about 1.45 greater than on Earth even though the gravity of Titan is only about 10% of Earth's. The atmosphere is very cold. At the surface the temperature is only 94 kelvins, so many complex and fragile molecules can form and persist. The high-altitude hazes also prevent much ultraviolet light from reaching the surface, which also reduces the photodissociation of molecules.

The Cassini/Huygens spacecraft arrived at Saturn in 2004, and returned the first detailed images of Titan's surface using a radar system. The Huygens lander also parachuted through the atmosphere taking pictures and making measurements along the way, and landed on the surface. Near Titan's south pole, an enigmatic dark feature named Ontario Lacus was identified (and later confirmed to be a lake). A possible shoreline was also identified near the pole via radar imagery.

Molecule		Abundance	On Earth
Nitrogen	N2	97%	78%
Oxygen	O2	0	21%
Argon	Ar	6%	0.93%
Methane	CH4	5%	0.00018 %
Hydrogen	H2	0.2 %	0.000055 %
Ethane	C2H6	10 ppm	0.0005 ppm
Carbon monoxide	CO	10 ppm	0.1 - 0.3 ppm
Acetylene	C2H2	2 ppm	0.0001 ppm
Propane	C3H8	500 ppb	
Hydrogen cyanide	HCN	170 ppb	
Ethylene	C2H4	100 ppb	
Acetonitrile	CH3CN	5 ppb	
Carbon dioxide	CO2	10 ppb	400000 ppb
Cyanoacetylene	HC3N	10 ppb	
Methylacetylene	CH3C2H	5 ppb	
Cyanogen	C2N2	5 ppb	
Water vapor	H2O	8 ppb	< 5%
Diacetylene	C4H2	1 ppb	

Note: 100 ppm = 0.01%

Based on the observations, definitive evidence of lakes filled with methane on Saturn's moon Titan was announced in January 2007. Overall, the Cassini radar observations have shown that lakes cover only a few percent of the surface, making Titan much drier than Earth. Early radar measurements indicated that Ontario Lacus was extremely shallow, with an average depth of 0.4–3 m, and a maximum depth of 3 to 7 m (9.8 to 23.0 ft).

In contrast, the northern hemisphere's Ligeia Mare was initially mapped to depths exceeding 8 m, the maximum discernable by the radar instrument and the analysis techniques of the time. Later science analysis, released in 2014, more fully mapped the depths of

Titan's three methane seas and showed depths of more than 200 meters (660 ft).

Ligeia Mare averages from 20 to 40 m (66 to 131 ft) in depth, while other parts of Ligeia did not register any radar reflection at all, indicating a depth of more than 200 m (660 ft). While only the second largest of Titan's methane seas, Ligeia "contains enough liquid methane to fill three Lake Michigans."

The density of Titan is consistent with a body that is about 60% rock and 40% water. Titan's surface can rise and fall by up to 10 meters during each orbit. The most likely model of Titan is one in which an icy shell dozens of kilometers thick floats atop a global ocean. Titan's ocean may lie no more than 100 kilometers (62 mi) below its surface.

The Huygens probe landed just off the easternmost tip of a bright region now called Adiri, and photographed a dark plain covered in small rocks and pebbles, which are composed of water ice. The two rocks just below the middle of the image on the right are smaller than they may appear: the left-hand one is 15 centimeters across, and the one in the center is 4 centimeters across, at a distance of about 85 centimeters from Huygens. (Image credit: NASA/ESA/Cassini/Huygens)

Interplanetary Travel

There have been several conceptual missions proposed in recent years for returning a robotic space probe to Titan. Initial conceptual work has been completed for such missions by NASA, the ESA and JPL. At present, none of these proposals have become funded missions.

The Titan Saturn System Mission (TSSM) was a joint NASA/ESA proposal for exploration of Saturn's moons. It envisions a hot-air balloon floating in Titan's atmosphere for six months. The Titan Mare Explorer (TiME) would be a low-cost lander that would splash down in a lake in Titan's northern hemisphere and float on the surface of the lake for 3 to 6 months.

The Aerial Vehicle for In-situ and Airborne Titan Reconnaissance (AVIATR) would be an unmanned plane (or drone) that would fly through Titan's atmosphere and take high-definition images of the surface of Titan. The Titan Lake In-situ Sampling Propelled Explorer (TALISE). The major difference compared to the TiME probe would be that TALISE is envisioned with its own propulsion system and would therefore not be limited to simply drifting on the lake when it splashes down. Finally, a NASAA Discovery Program contestant for its mission is Journey to Enceladus and Titan (JET), an astrobiology Saturn orbiter that would assess the habitability potential of Enceladus and Titan.

The Moons of Uranus

All major moons comprise approximately equal amounts rock and ice, except Miranda, which is made primarily of ice. The ice component may include ammonia and carbon dioxide. Their surfaces are heavily cratered, though all of them (except Umbriel) show signs of resurfacing and in the case of Miranda, race-track like structures called coronae possibly caused by upwelling material. Ariel appears to have the youngest surface with the fewest impact craters, while Umbriel's appears oldest. The largest Uranian moons

136

may be internally differentiated, with rocky cores at their centers surrounded by ice mantles. Titania and Oberon may harbor liquid water oceans at the core/mantle boundary. The major moons of Uranus are airless bodies. For instance, Titania was shown to possess no atmosphere at a pressure larger than 10–20 nanobar.

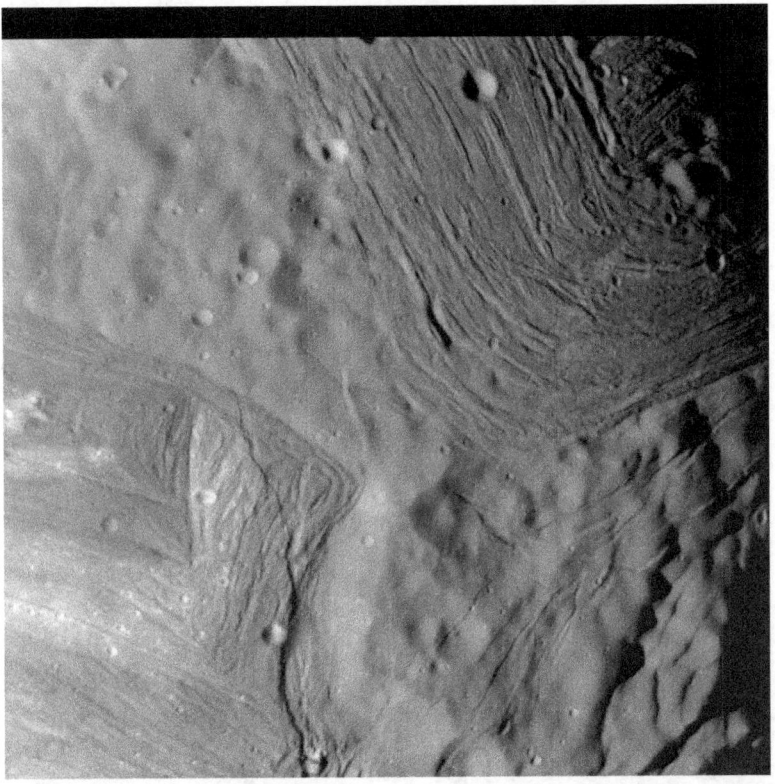

Complex surface of Miranda seen by Voyager 2 spacecraft. Details as small as 600 meters across can be seen. The area shown is about 150 kilometers (93 miles) on a side. (Image credit: NASA/JPL)

The Moons of Neptune

Triton, the largest moon of the planet Neptune, has a surface of mostly frozen nitrogen, a mostly water ice crust, an icy mantle and a substantial core of rock and metal. The core makes up two-thirds of its total mass. 55% of Triton's surface is covered with frozen nitrogen, with water ice comprising 15–35% and dry ice (frozen carbon dioxide) forming the remaining 10–20%. Trace ices include 0.1% methane and 0.05% carbon monoxide. There could also be ammonia on the surface. Triton's density implies it is probably about 30–45% water ice, with the remainder being rocky material. Triton's reddish color is thought to be the result of methane ice, which is converted to tholins under bombardment from ultraviolet radiation.

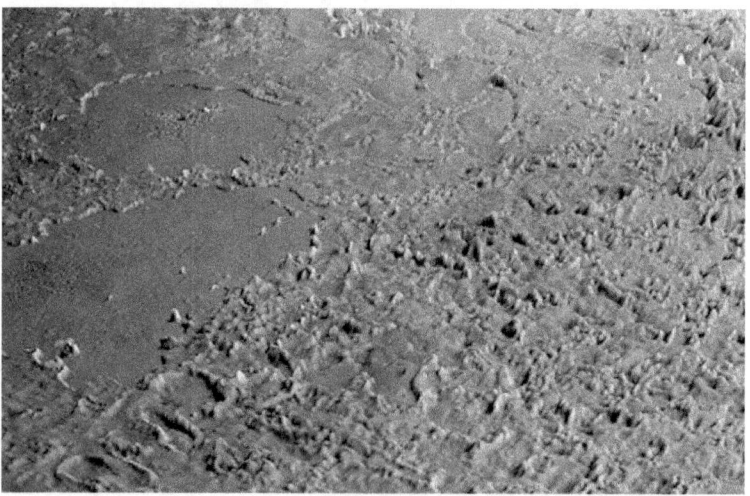

In this Voyager 2 image, the rugged terrain in the foreground is Triton's infamous cantaloupe terrain, most likely formed when the icy crust of Triton underwent wholesale overturn, forming large numbers of rising blobs of ice (diapirs). The numerous irregular mounds are a few hundred meters (several hundred feet) high and a few kilometers (several miles) across and formed when the top of the crust buckled during overturn. The large walled plains are of unknown origin, although the irregular pit in the center of the background walled plain may be volcanic in nature. These plains are approximately 150 meters (0.093 miles) deep and 200 to 250 kilometers (124 to 155 miles) across. The surface of Triton is very rugged, scarred by rising blobs of ice, faults and volcanic pits and lava flows composed of water and other ices. The surface is also extremely young and sparsely cratered, and could be geologically active today. This scene is on the order of 500 kilometers (310 miles) across. Vertical relief has been exaggerated by a factor of 25 to aid interpretation. (Image credit: NASA/JPL/Universities Space Research Association/Lunar & Planetary Institute)

Water comprises Triton's mantle, which lies over a core of rock and metal. There is enough rock in Triton's interior for radioactive decay to power convection in the mantle. The heat may even be sufficient to maintain a subterranean ocean similar to that which is hypothesized to exist underneath the surface of Europa. If present, a layer of liquid water would suggest the possibility of life. Triton has a tenuous nitrogen atmosphere, with trace amounts of carbon monoxide and small amounts of methane near the surface. The atmosphere of Triton is thought to have resulted from evaporation of nitrogen from its surface.

Streaks on Triton's surface left by geyser plumes suggest that the troposphere is driven by seasonal winds capable of moving material of over a micrometer in size. A haze permeates most of Triton's troposphere, thought to be composed largely of hydrocarbons and nitriles created by the action of sunlight on methane. Triton's atmosphere also possesses clouds of condensed nitrogen that lie between 1 and 3 km from the surface.

Pluto and Beyond

Pluto is one of the most famous members of the Kuiper Belt located beyond the orbit of Neptune. Recently demoted by the International Astronomical Union from a planet to a dwarf planet in 2006, it is now viewed as merely one of hundreds of similar though smaller objects that have been cataloged out in the most remote depths of our solar system. Prior to the July 2015 flyby of Pluto by NASA's New Horizon spacecraft, little was known about Pluto other than its apparently ice-bound surface and atmosphere, which periodically freezes to the surface on the winter, or vaporizes in the summer. Hubble studies suggested a complicated, blotchy surface, and a retinue of moons including Charon. At this distance, sunlight is 900 times fainter than on Earth so noontime on Pluto looks like Earth twilight. The time taken by a light signal to reach Pluto from Earth is about 5 hours, so any human or robotic mining activity will have to be largely autonomous. Given that the surface is ice-bound, this is the most expensive reservoir of minable water ice in the entire solar system!

The highest resolution image of Pluto's surface captured by New Horizons reveals few craters and evidence of ice flows of apparently recent origins. The high average density of Pluto suggests that as much as 70% of its interior is a rocky core, surrounded by an icy mantle extending to the surface. Between the core and mantle, there may exist a liquid ocean layer perhaps 100-km thick.

The fact that Pluto is a Double Planet system with its large moon Charon and resembles pour own earth-Moon system suggests that it may have formed in the same was as our moon via a giant impact over 4 billion years ago. The impact liquefied the outer ice mantle of Pluto and scattered some of it into space to form Charon.

Over eons the surface of Pluto re-solidified, but that a remnant of its liquid state remains in the core-mantle boundary. This could explain why few craters exist, because the crust remained very active for quite a while after Charon's formation, destroying any craters that might have formed thereafter.

Clearly, Pluto and Charon are rich in water ice, nitrogen and methane, but all traces of rocky materials have long since settled to their core regions and are inaccessible from the surface. This precludes any mining activities, although as for other ice-bound moons, the material can be used for rocket fuel.

141

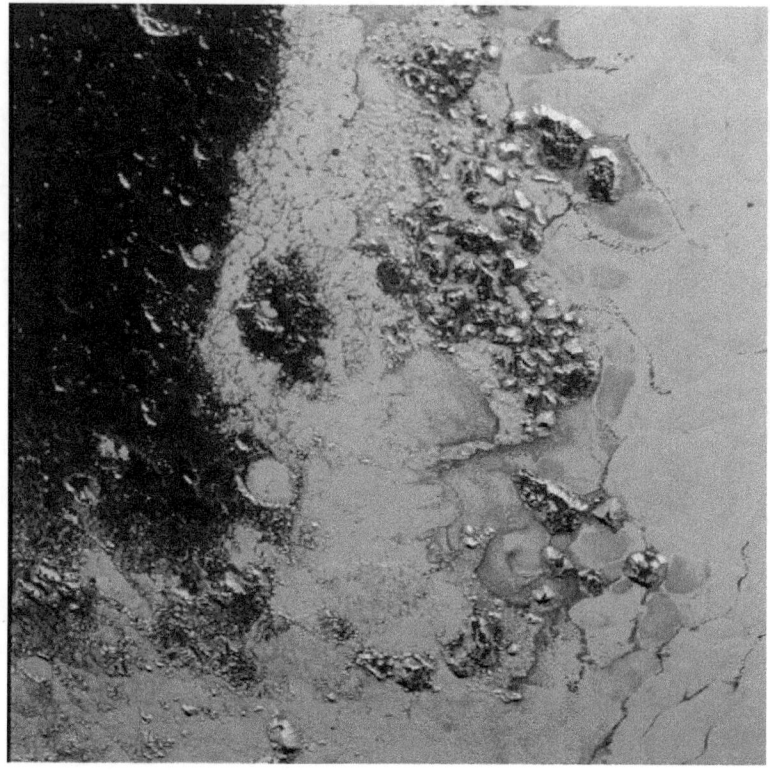

Comet Mining

Among the hundreds of thousands of small bodies in the solar system, there are many resources that could, and probably will be, be retrieved, even from comets like C/1999 S4.

Two fragments from the disintegrating comet 73P/Schwassmann-Wachmann 3 are very similar in icy composition, which shows this comet is strongly depleted in CH_3OH, H_2CO, C_2H_2, C_2H_6, NH_3, and H_2S while normal in H_CN, CH_3CN, $HNCO$ and CS contents relative to water.

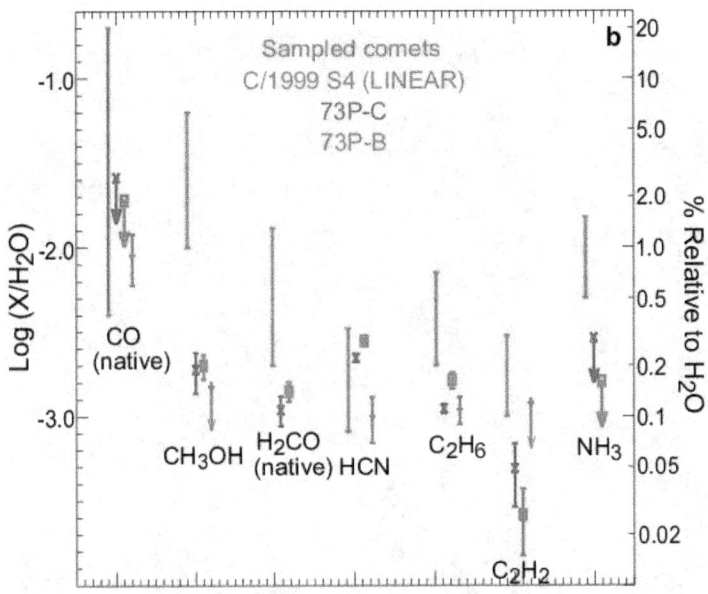

Comet C/2009 P1 is a long period Oort Cloud comet that is rich in carbon monoxide, which is one of the most abundant ice molecules in interstellar dust clouds, but is absent in inner-system comets. It is also rich in volatile compounds that are usually absent from inner-system comets.

Molecule	Formula	Abundance (Water=100%)
Water	H_2O	100%
Ethane	C_2H_6	2.9%
Methane	CH_4	1.34%
Carbon monoxide	CO	13.45%
Cyanide	HCN	0.37%
Acetylene	C_2H_2	0.14%
Ammonia	NH_3	1.69%
Carbonyl sulfide	OCS	0.20%
Hydrogen deuterium oxide	HDO	0.9%
Methanol	CH_3OH	3.9%

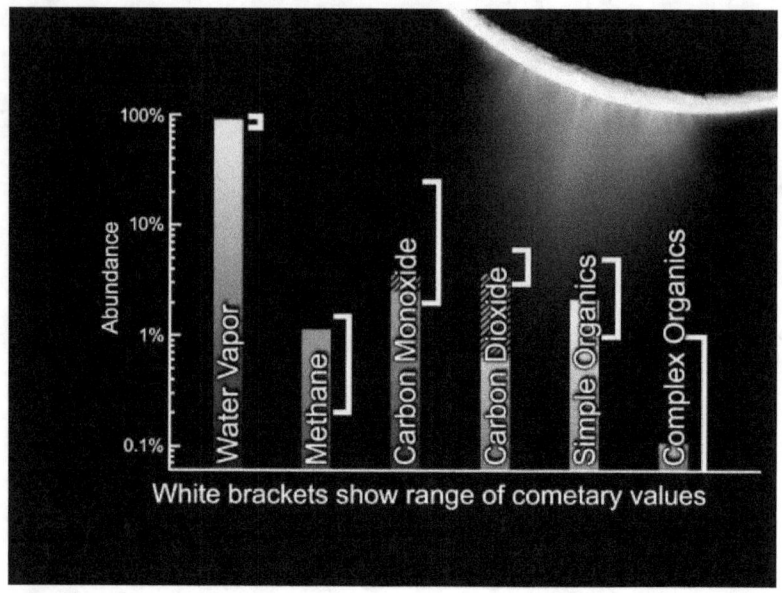

Enceladus's plume was found to have a comet-like chemistry by Cassini's Ion and Neutral Mass Spectrometer during its fly-through of the plume on 12 March 2008. Water vapor, methane, carbon monoxide, carbon dioxide, simple organics and complex organics were identified in the Enceladus plume. The bar graph shows the chemical constituents in percentage of abundance found in comets compared to those found in Enceladus's plume. (Image Credit: NASA/JPL/SwRI)

This table shows that, if you extract 100 kilograms of water from an average comet nucleus, you may also get 0.1x100 kg = 100 grams of methyl formate mixed into this sample. Because most of these ingredients are poisonous to humans, it would be a fatal act to simply melt comet 'ice' and drink it! Astronauts will need to process comet ice to remove all of its impurities to yield what will basically be distilled water with no minerals. The other components could then be used to extract hydrogen and oxygen for fuel.

Molecule		Relative Abundance
Water	(H_2O)	100
Carbon monoxide	(CO)	23
Carbon dioxide	(CO_2)	6
Methanol	(CH_3OH)	2.4
Hydrogen sulfide	(H_2S)	1.5
Formaldehyde	(H_2CO)	1.1
Ammonia	(NH_3)	0.7
Methane	(CH_4)	0.6
Carbonyl sulfide	(OCS)	0.4
Ethane	(C_2H_6)	0.3
Sulfur monoxide	(SO)	0.3
Hydrogen cyanide	(HCN)	0.3
Carbon disulfide	(CS_2)	0.2
Carbon monosulfide	(CS)	0.2
Acetylene	(C_2H_2)	0.1
Formic acid	(HCOOH)	0.1
Isocyanic acid	(HNCO)	0.1
Methyl formate	($HCOOCH_3$)	0.1

The NASA Stardust spacecraft in January 2006 recovered samples of comet dust from the comet Wild 2, which appear to be weakly constructed mixtures of very small grains with a few larger grains. Also, a wide range of high- and low-temperature minerals, from olivine to low- and high-calcium pyroxene compositions, is present in the Wild 2 samples.

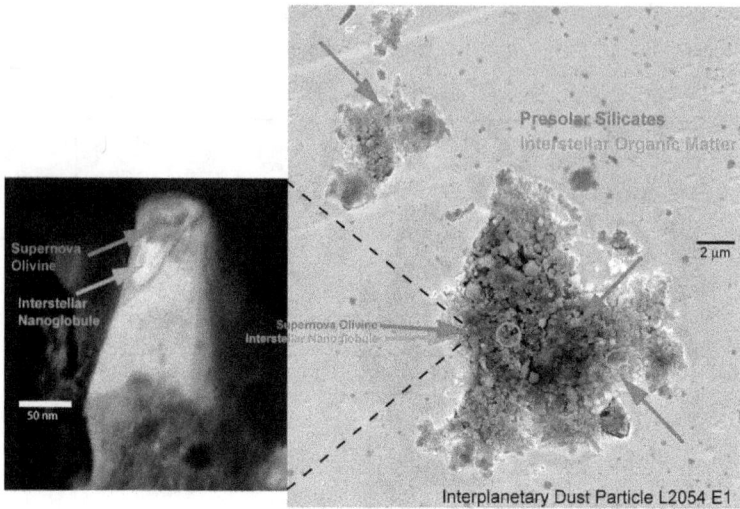

Interplanetary Dust Particle L2054 E1

The samples are predominantly fine-grained, loosely bound aggregates, most also containing much larger individual crystals of olivine, pyroxene and iron/nickel sulfides. All analyses suggest that small and large Wild 2 particles are composed of a similar, if not identical, suite of minerals.

Comet 67P/Churyumov-Gerasimenko

67P/Churyumov–Gerasimenko is a comet, originally from the Kuiper belt, with a current orbital period of 6.45 years, a rotation period of approximately 12.4 hours and a maximum velocity of 135,000 km/h (38 km/s; 84,000 mph). Churyumov–Gerasimenko is approximately 4.3 by 4.1 km (2.7 by 2.5 mi) at its longest and widest dimensions.

On 6 June 2014, water vapor was detected being released at a rate of roughly 1 liter/s (0.26 USgal/s) when Rosetta was 360,000 km (220,000 mi) from Churyumov–Gerasimenko and 3.9 AU (580,000,000 km) from the Sun. On 14 July 2014, images taken by Rosetta showed that its nucleus is irregular in shape with two distinct lobes. One explanation is that it is a contact binary formed by low-speed accretion between two comets, but it may instead have resulted from asymmetric erosion due to ice sublimating from its surface to leave behind its lobed shape. The size of the nucleus is estimated to be 3.5×4 km (2.2×2.5 mi).

Comet 67P/Churyumov-Gerasimenko by Rosetta's OSIRIS narrow-angle camera on August 3, 2014 from a distance of 285 km. The image resolution is 5.3 meters/pixel. (Credit: ESA/Rosetta/MPS)

Comet 67P is classed as a dusty comet, with a dust to gas emission ratio of approximately 2:1. The peak dust production rate in 2002/03 was estimated at approximately 60 kg per second, although values as high as 220 kg per second were reported in 1982/83.

147

The surface composition of this comet has been deduced from Rosetta instruments. The VIRTIS instrument measured sunlight reflected from the comet's surface and the data showed that the surface is rich in polyaromatic organic solids mixed with sulfites and iron nickel alloys. The presence of the sulfides, carbon-hydrogen groups and OH chemical groups in the spectra also imply the surface has these components.

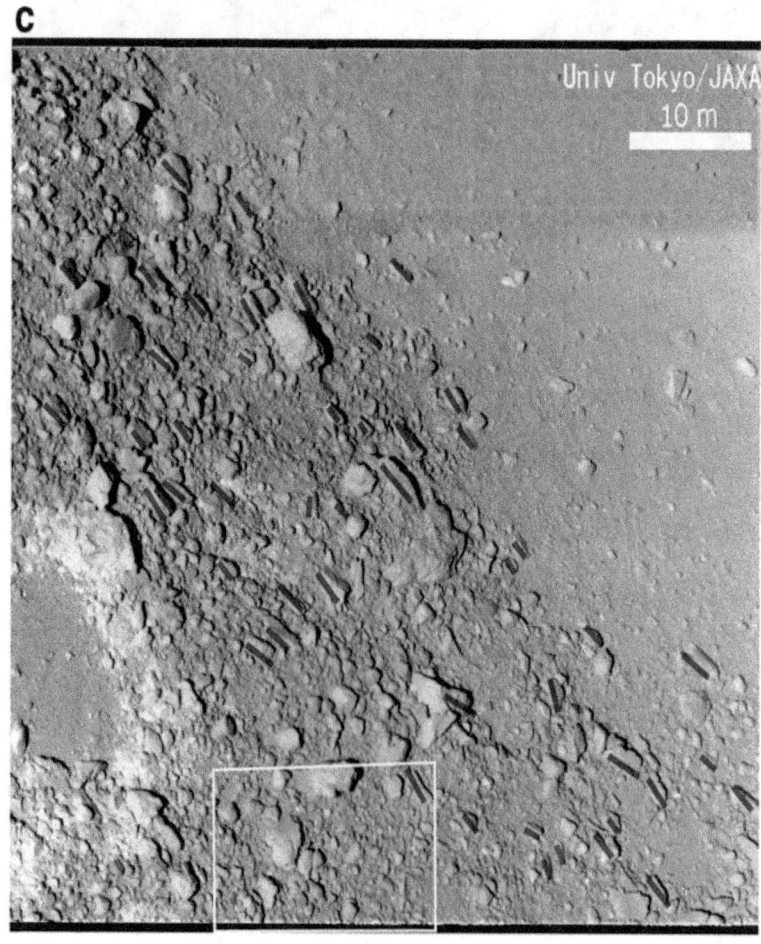

Tourism

Most science fiction stories assume that interplanetary travel will have a significant tourism component to it. But first we have to crawl before we can walk. In the past, getting off of Earth has required huge launch vehicles largely built under government contract for NASA and the Department of Defense. The major launch vehicle providers for the government also had side-businesses to launch commercial communications or mapping satellites into space, but no humans. After the Space Shuttle was mothballed, NASA opened up the possibility of providing domestic companies with the contracts to ferry resources and humans to the International Space Station. Very quickly, new companies began to rise to this challenge, largely bankrolled by a handful of impressive individuals.

Space Exploration Technologies Corporation (SpaceX) is an American aerospace manufacturer and space transport services company with its headquarters in Hawthorne, California, USA, founded in 2002 by former PayPal entrepreneur and Tesla Motors CEO Elon Musk. Its goal was reducing space transportation costs to enable the colonization of Mars. It has developed the Falcon 1 and Falcon 9 launch vehicles, both of which were designed from conception to eventually become reusable, and the Dragon spacecraft which is flown into orbit by the Falcon 9 launch vehicle to supply the International Space Station with cargo. A manned version of Dragon is in development.

Interplanetary Travel

SpaceX's achievements include the first privately funded, liquid-propellant rocket (Falcon 1) to reach orbit on 28 September 2008; the first privately funded company to successfully launch, orbit and recover a spacecraft (Dragon) on 9 December 2010; and the first private company to send a spacecraft (Dragon) to the International Space Station on 25 May 2012. The launch of SES-8, on 3 December 2013, was the first SpaceX delivery into geosynchronous orbit.

In 2006, NASA awarded SpaceX a Commercial Orbital Transportation Services (COTS) contract to design and demonstrate a launch system to resupply cargo to the International Space Station (ISS). SpaceX, as of May 2015 has flown six missions to the ISS under a cargo resupply contract. NASA has also awarded SpaceX a contract to develop and demonstrate a human-rated Dragon as part of its Commercial Crew Development (CCDev) program to transport crew to the ISS.

Musk has stated that one of his goals is to improve the cost and reliability of access to space, ultimately by a factor of ten. The company plans in 2004 called for "development of a heavy lift product and even a super-heavy, if there is customer demand" with each size increase resulting in a significant decrease in cost per pound to orbit. Musk said: "I believe $500 per pound ($1,100/kg) or less is very achievable."

In June 2013, Musk used the descriptor Mars Colonial Transporter to refer to the privately funded development project to design and build a spaceflight system of rocket engines, launch vehicles and space capsules to transport humans to Mars and return to Earth. Once the Falcon Heavy and Dragon v2 crew version are flying, the focus for the company engineering team will be on developing the technology to support the transport infrastructure necessary for Mars missions.

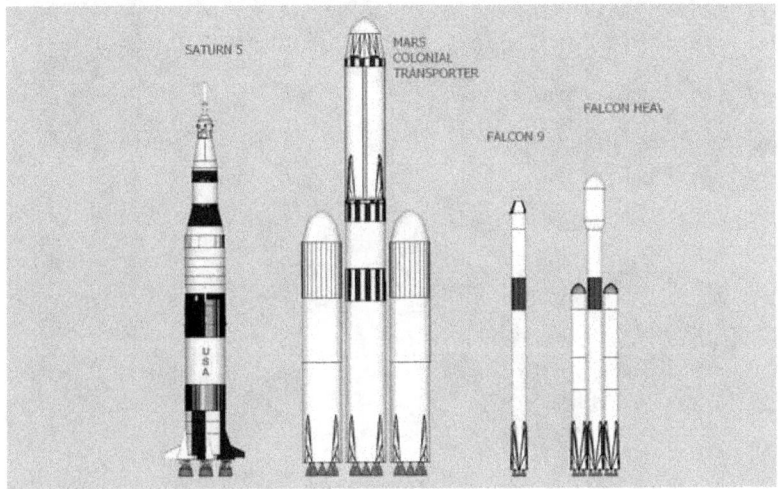

On September 16, 2014, NASA chose SpaceX and Boeing as the two companies that will be funded to develop systems to transport U.S. crews to and from the space station. SpaceX won $2.6B to complete and certify Crew Dragon spacecraft by 2017. (Boeing won $4.2B to complete and certify their CST-100.) The contracts include at least one crewed flight test with at least one NASA astronaut aboard. Once Crew Dragon achieves NASA certification, the contract requires SpaceX to conduct at least two, and as many as six, crewed missions to the space station.

SpaceX has also begun the design of a super-heavy-lift launch vehicle that will consist of one or three 10-meter (33 ft)-diameter cores and use nine Raptor LOX/methane engines to power each core. The MCT launch vehicle is also intended to be reusable and will produce approximately 40 or 120 meganewtons (9,000,000 or 27,000,000 lbf) of thrust at liftoff. Development of the Mars Colonial Transporter and its super-heavy launch vehicle will be the major focus of SpaceX once Falcon Heavy and DragonCrew are flying regularly

Virgin Galactic is an American-based spaceflight company within the British Virgin Group. It is developing commercial spacecraft and aims to provide suborbital spaceflights to space tourists, suborbital launches for space science missions, and orbital launches of small satellites. Further in the future, Virgin Galactic plans to provide orbital human spaceflights as well. The company also hopes to develop an orbital launch vehicle. SpaceShipTwo, Virgin Galactic's suborbital spacecraft, is air launched from beneath a carrier airplane known as White Knight Two.

To date, the three test flights of the SS2 have only reached an altitude of around 71,000 ft, approximately 13 miles; in order to receive a Federal Aviation Administration license to carry passengers, the craft needs to complete test missions at full speed and 62-mile height. Following the announcement of further delays, UK newspaper The Sunday Times reported that Branson faced a backlash from those who had booked flights with Virgin Galactic, with the company having received $80 million in fares and deposits.

Tom Bower, author of *Branson: The Man behind the Mask*, told the Sunday Times: "They spent 10 years trying to perfect one engine and failed. They are now trying to use a different engine and get into space in six months. It's just not feasible." BBC science editor David Shukman commented in October 2014, that "[Branson's] enthusiasm and determination [are] undoubted. But his most recent promises of launching the first passenger trip by the end of this year had already started to look unrealistic some months ago." In

February 2007, Virgin announced that they had signed a memorandum of understanding with NASA to explore the potential for collaboration, but, to date, this has produced only a relatively small contract in 2011 of up to $4.5 million for research flights.

Blue Origin is an American privately funded aerospace manufacturer set up by Amazon.com founder Jeff Bezos. The company is developing technologies to enable private human access to space with the goal of dramatically lower cost and increased reliability. It is employing an incremental approach from suborbital to orbital flight, with each developmental step building on its prior work. Blue Origin is developing a variety of technologies, with a focus on rocket-powered Vertical Takeoff and Vertical Landing (VTVL) vehicles for access to suborbital and orbital space.

Initially focused on sub-orbital spaceflight, the company has built and flown a testbed of its New Shepard spacecraft design at their

Culberson County, Texas facility. The first developmental test flight of the New Shepard was April 29, 2015. The uncrewed vehicle flew to its planned test altitude of more than 307,000 feet (93,500 meters) and achieved a top speed of Mach 3.

As of July 2014, Bezos has invested over $500 million of his money into Blue Origin. In September 2014, the company and United Launch Alliance (ULA) entered into a partnership where Blue will produce a large rocket engine—the BE-4—for the successor to the Atlas V, a 10,000–19,000 kilograms (22,000–42,000 lb)-class launch vehicle that has launched US national security payloads since the early 2000s. The announcement included that Blue had been working on this engine for three years prior to the public announcement, and that the first flight on the new rocket could be as early as 2019. In April 2015, Blue Origin announced that it had completed acceptance testing of the BE-3 engine that will power the New Shepard space capsule that will be used for Blue Origin suborbital flights.

Bigelow Aerospace is an American space technology startup company, based in North Las Vegas, Nevada that is pioneering work on expandable space station modules. Bigelow Aerospace was founded by Robert Bigelow in 1998. and is funded in large part by the fortune Bigelow gained through his ownership of the hotel chain Budget Suites of America. By 2013, Bigelow had invested $250 million in the company. Bigelow has stated on multiple occasions that he is prepared to fund Bigelow Aerospace with about $500 million through 2015 in order to achieve launch of full-scale hardware. Bigelow is pioneering a new market in a flexible and configurable set of space habitats. Moreover, industry observers have noted that Bigelow is demonstrating audacity to pioneer such a market "in a capital-intensive, highly-regulated industry like spaceflight."

Bigelow Aerospace anticipates that its inflatable modules will be more durable than rigid modules. In ground-based testing, micrometeoroids capable of puncturing standard ISS module materials penetrated only about half-way through the Bigelow skin.

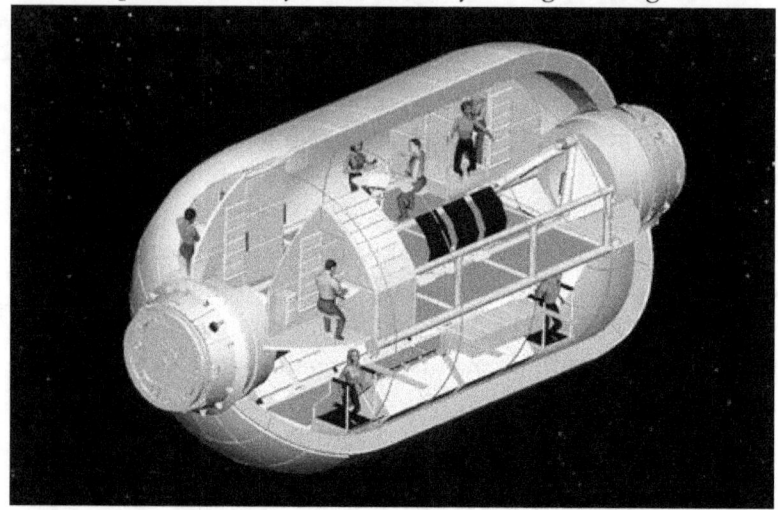

In December 2012, Bigelow began development work on Bigelow Expandable Activity Module (BEAM) under a $17.8 million NASA contract. In 2015, BEAM is projected to be transported to ISS inside the unpressurized cargo trunk of a SpaceX Dragon during the SpaceX CRS-8 cargo mission. The spaceflight is intended to test the BEAM module's structural integrity, leak rate, radiation dosage and temperature changes over a notional two-year-long mission. At the end of BEAM's mission, the module is planned to be removed from the ISS and burn up during reentry.

The Bigelow Next-Generation Commercial Space Station is a private orbital space complex currently under development by Bigelow. The space station will include both Sundancer and BA 330 expandable spacecraft modules and a central docking node, propulsion, solar arrays, and attached crew capsules. Bigelow has publicly shown space station design configurations with up to nine BA 330 modules containing 100,000 cu ft (2,800 m3) of habitable space. Bigelow began to publicly refer to the initial configuration — two Sundancer modules and one BA 330 module — of the first Bigelow station as "Space Complex Alpha" in October 2010. A second orbital station, Space Complex Bravo, is scheduled to begin launches in 2016.

© Bigelow Aerospace 2013

In December 2014, the FAA Office of Commercial Space Transportation (AST) completed a review of the proposed Bigelow lunar habitat, and indicated that "it was willing to use its authority to ensure Bigelow could carry out its [lunar] activities ... without interference from other [US] companies licensed by the FAA" [and that the FAA would] use its launch licensing authority, as best it can, to protect private sector assets on the Moon and to provide a safe environment for companies to conduct peaceful commercial activities without fear of harmful interference from other AST licensees."

157

Interorbital Systems Corporation (IOS) is an American aerospace manufacturer based in Mojave, California. It was founded in 1996 by Roderick and Randa Milliron, who also co-founded Trans Lunar Research. Interorbital Systems is currently working on a line of launch vehicles aimed at winning the Google Lunar X Prize. The company was also a competitor for both the Ansari X Prize and America's Space Prize. In 2007, Interorbital Systems was pursuing two separate lines of research. The first is Sea Star TSAAHTO, a small rocket capable of delivering small satellite payloads into orbit, and so generate capital for pursuing the companies flagship program: the Neptune; a line of launch vehicles designed to put humans into orbit.

XCOR Aerospace is an American private rocket engine and spaceflight development company based at the Mojave Air and Space Port in Mojave, California. XCOR was formed by former members of the Rotary Rocket rocket engine development team in September, 1999. XCOR is headed by Jeff Greason, who is the CEO. The Lynx is capable of carrying a pilot and a passenger or payload on sub-orbital spaceflights over 100 kilometers (62 mi). Between 20 and 50 test flights of Lynx were planned, along with numerous static engine firings on the ground. A full step-by-step set of taxi tests, runway hops and full-up flights were planned to get the vehicle to a state of operational readiness. Lynx was envisaged to be roughly the size of a small private airplane. It would be capable of flying several times a day making use of reusable, non-toxic engines to help keep the space plane's operating costs low. The Lynx superseded a previous design, the Xerus spaceplane. The Lynx was initially announced on March 26, 2008, with plans for an operational vehicle within two years. That date slipped, first to early 2012, and then to 2015. The Mark II would fly twelve to eighteen months afterwards depending on how fast the prototype moved through the test program. As of July 2012, XCOR had presold 175 Lynx flights at $95,000 each.

When should we go?

In all of human history, there has never been a 'good time' to go exploring. If past is prolog, we tend not to wait for all of Society's issues to finally be in order before we launch an expedition. Columbus did not wait until poverty had been solved in Europe before striking out across the Atlantic, nor did the Vikings wait until all was Peace and Harmony at home before venturing to Greenland and North America. The fact of the matter is, a journey of exploration begins just about any time one can gather together the required resources and a few daring travelers. What is certainly true is that there are better times to try to do this than others.

Before we had the daring-do of the Space Age, we had in the previous centuries the great expeditions into the African continent, and to the polar regions that attracted not only explorers for the historical glory, but countries vying for the prize of political one-upmanship. Contrary to today's risk-averse societies, back then explorers routinely died as they took risks exploring completely unfamiliar areas. The Age of Discovery began in the early-15th Century with the Atlantic archipelagos and African coastline, through the initial voyages to North America by Columbus.

By the 16th century, all major human civilizations slowly introduced each other, and the first detailed maps of the entire globe were created. The late-18th century brought the fabled James Cook Expeditions into the Pacific Ocean, and those of Alexander von Humboldt into Africa. By the 19th century, the Alfred Wallace expedition into the Amazon and South America were followed by the Arctic and Antarctic expeditions and the search for the Northwest Passage across northern Canada.

The last major terrestrial expeditions were to study the Antarctic regions, and the high-speed trek to the pole by Norwegian Roal Amundson in 1911 was the capstone to the end of an era of exploration. By this time, there were essentially no major unknown islands, continents or places to visit on the 29% of Earth's surface that was solid ground. There would still be deep caves to attract spelunkers, and geographically minute jungles areas in the Amazon or Africa that could still use cataloging, but that was about it.

Meanwhile, there was a huge landmass covered by Earth's oceans that still beckoned. In favorable places near the shallower coastlines and continental margins, archeologists and treasure hunters concentrated their work. In the greater depths, military intelligence assembled enormously detailed maps of the untouchable, hidden landscapes, first using sonar, but then after the 1950's using radar. By now, we have a complete map of Earth's entire surface, but as for the vast oceans prowled by military, commercial and scientific vessels, there is still no compelling need to physically stand on the ocean's bottom. Robotic exploration works just fine. The occasional scientists that make the hazardous deep-sea journey are rewarded by discoveries worthy of a just a few professional journal articles. (Image credit: National Geographic)

Once the chapter closed on humanity's epic work of mapping our entire Earth, other efforts began to map its deep interior through 3-D seismic tomography. (Image credit: Ebru Bozdağ, University of Nice Sophia Antipolis, and David Pugmire, Oak Ridge National Laboratory)

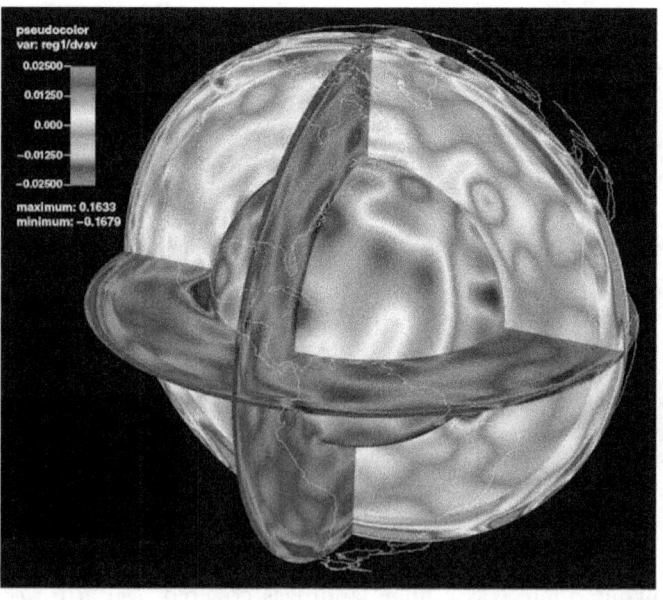

This process has only just begun, but thanks to the thousands of well-placed seismographic stations around the globe, and powerful supercomputer-based mathematical analysis, we are slowly discovering the complex and changing world beneath our feet, and more importantly, below our earthquake-prone cities. There are few who would say that this investment is being squandered for mere scientific curiosity. Like the expeditions of the previous centuries, scientific curiosity has been perfectly melded with economic and public safety necessity to allow robust funding of these projects every year.

In all these investigations, the issue of 'when should we start' was at first driven by scientific curiosity, then as the practicality of the investment was demonstrated, either for military, resource or public health reasons, the research continued and even escalated. Surface exploration is still supported as a possible way to locate new industrial resources through geologic or fossil fuel discoveries, but satellite surveys and areal magnetometer studies are just as effective and less expensive than putting people on the ground. The oceans still remain a potential reservoir of new resources although generally very hazardous and expensive to extract.

The problem with space exploration is, of course, that it is a very costly enterprise. In the past, a few hardened travelers could get on a horse with minimal supplies and 'discover' an entire continent, or equip a few ships to sail the seas and discover distant lands. Exploration was funded by literally pocket change, and generally did not involve much more than finding a few wealthy benefactors to front the money to purchase the resources. This is not at all the case for reaching space, or traveling to distant planets and planetoids. Launch vehicles, not horses and sails, are needed to cover a mere 200 miles to Low Earth Orbit (LEO), and these can be very costly depending on whether they just transport material or humans.

Nevertheless, supported by national treasuries, federal budgets and far-sighted corporate investments, the advent of the Space Age has allowed us to rapidly commercialize space out to geosynchronous orbit (66,000 km). There is also an enormous strategic value to space, especially LEO where the clearest images of Earth can be obtained with the smallest optical systems. So the political need to be in space at a particular time in the 1950s and 1960s has now matured into a major commercial and military necessity. We embarked on space exploration in the late 1950s for the simple political reason of demonstrating the technological superiority of one nation over another. The public outcry that 'we need to spend more time curing poverty rather than exploring space and walking on the moon' was a common outcry during these early years. It was clearly not the right time to embark on exploring space, and the focusing of up to 4% of the US annual budget on the race to the moon was a huge extravagance at the time. Arguably, had we not done it then, the international area for US science and leadership would be very different today. The huge, high-tech stimulus to the US economy in the 1960s that resulted in our arriving at the moon as the new generation of explorers, was an investment returned with the spectacular commercialization of space during the 1970s and 1980s.

The lesson learned was, not that a particular device or system used in the moon landing was then commercialized for consumption by the public, but that the technological base that made these devices possible could be easily re-tooled to give us CT scanners, the internet, and cell phones with GPS only a few decades later. This is the sub-text for why exploration is vital to a society. Exploration forces us to adapt, and develop new technologies to meet new conditions and situations. Without the challenge of solving new problems, we limit ourselves to older ways of doing business and science. We could never have gone to the moon with vacuum tube technology, nor would cellular phones today be as we find them.

So, the right time to go exploring is always Right Now. It is never an activity that we dare to postpone to the future when circumstances may be more favorable. The truth is, voyages of

164

discovery will never be cheaper than they are now. We went to the moon in 1969 following a 6-year development process costing a few tens of billions of dollars. We can do this same trip today, but at much higher cost, and we would have lost 50 years of US leadership in advanced technology development along the way. We would then have far more to complain about today than the collapse of the industrial 'Iron Belt'.

Perhaps the best clues to what the immediate future might bring can be found in the official comments by various international space agency representatives.

In a major space policy speech at Kennedy Space Center on April 15, 2010, U.S. President Barack Obama predicted a manned Mars mission to orbit the planet by the mid-2030s, followed by a landing: "*By the mid-2030s, I believe we can send humans to orbit Mars and return them safely to Earth. And a landing on Mars will follow. And I expect to be around to see it.*" The United States Congress has mostly approved a new direction for NASA that includes canceling Bush's planned return to the Moon by 2020 and instead proposes asteroid exploration in 2025 (Asteroid Redirect Mission) and orbiting Mars in the 2030s.

The May 2-5, 2015 hearing in Washington DC called "*From Here to Mars,*" was a public presentation by NASA about its current plans for Mars in the 2030s. The precursor mission called the Asteroid Redirect Mission, plans to send a manned spacecraft to a near-earth asteroid, and then use robotic technology to knock the asteroid (or a boulder from it) into the moon's orbit. From there astronauts will be able to collect and study samples of the asteroid. The budget has this mission on target to happen by 2025.

William Gerstenmaier, a NASA administrator, presented the agency's goals before the Senate Subcommittee for Science and Space. "*Our architecture is designed for long-term human exploration of our*

Interplanetary Travel

solar system, including the goal of human missions to Mars". NASA is moving very quickly to develop the capabilities needed to send humans to an asteroid by 2025 and Mars in the 2030s. These are goals that have been outlined in the bipartisan NASA Authorization Act of 2010 and in the U.S. National Space Policy, also issued in 2010.

The ARM mission remains controversial because some complain that it is not actually a required mission for reaching Mars. The plan is that astronauts aboard the Orion spacecraft will explore a small Near Earth asteroid in the 2020s, returning to Earth with samples. This experience in human spaceflight beyond low-Earth orbit will help NASA test new systems and capabilities, such as Solar Electric Propulsion, which we'll need to send cargo as part of human missions to Mars. Meanwhile, beginning in FY 2018, NASA's powerful Space Launch System rocket will enable these "proving ground" missions to test new capabilities. Human missions to Mars will rely on Orion and an evolved version of SLS that will be the most powerful launch vehicle ever flown.

The real proving ground for Mars, though, is near the moon. It is there that NASA plans to spend a large part of the 20s, learning how to live and work in lunar Distant Retrograde Orbit, or DRO. Lunar DRO is a highly stable orbit where objects can remain steady for about a hundred years. At a June 19 briefing, Asteroid Redirect Mission Program Director Michelle Gates said NASA is currently interested in a lunar DRO with an altitude of about 75,000 kilometers. That's almost a fifth of the distance between the Earth and moon—a unique orbit unlike any humans have ever visited. The nice thing about it is that once you get from LEO to DRO, you have already invested 80% of the energy you need to get to Mars, making this a perfect orbit location near Earth to stockpile equipment and resources for Mars.

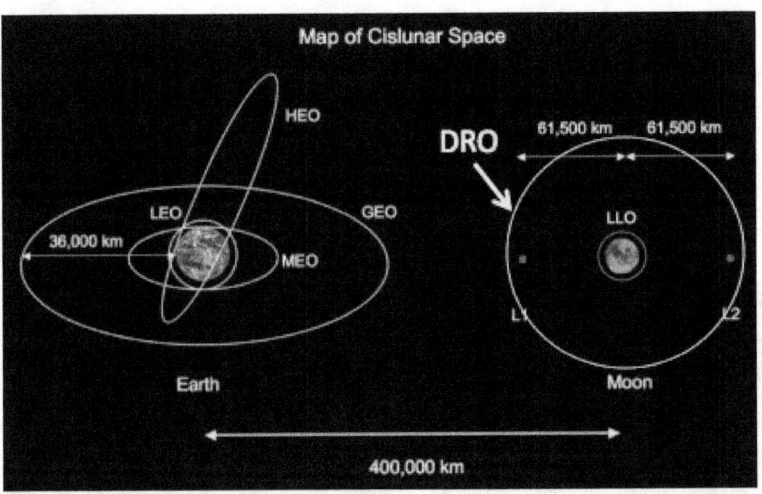

DRO orbit near moon (Image credit: Kirstyn Johnson/U. Colorado)

The idea for such a flight would be that at Kennedy Space Center, a crew of astronauts would climb into an Orion capsule and launch into space atop SLS. They would arrive at lunar DRO, dock with a habitat module, and depart for Mars. They would zip through interplanetary space as fast as they can, minimizing their exposure to radiation. In Martian orbit, they would meet up with more cargo and equipment. During the first trip, the crew might orbit without landing, Apollo 10-style. Another mission might land on Phobos. But eventually, a crew would plunge into the Martian atmosphere. Once on Mars, the crew would spend up to a year exploring the surface. They would return to Martian orbit and depart for Earth. Their transit habitat would get dropped off at lunar DRO for refurbishment, while the crew careened back into Earth's atmosphere aboard Orion and splashes down into the Pacific.

Meanwhile, as NASA is marshalling its meager funding resources for the major challenges of ARM and Mars, China's fledgling manned program has initiated a multistep space station program, sending the Tiangong 1, its first space lab and still-operating spacecraft, into orbit in September 2011. The liftoff of China's Tiangong 2 space lab, scheduled for 2016, is intended to sharpen China's space station construction skills. A Shenzhou 11 crewed spacecraft and a Tianzhou 1 cargo spacecraft would then be launched to dock with that facility. By about 2022, China's first space station would be fully operational.

Deciding what to do with the outer solar system after Cassini and New Horizons end their missions in 2017-2019 has not been easy. These destinations are very expensive and require missions costing upwards of $2 billion to accomplish the newer generation of science goals. Since 2012 both NASA and the European Space Agency (ESA) have tried to get funding for several mission concepts.

In one scenario, for $3 billion, NASA would build one orbiter, the Jupiter Europa, while ESA would provide the other: Jupiter Ganymede. The spacecraft would launch in 2020 from different spaceports with the goal of reaching Jupiter by 2026 and spending three years studying the planet and its moons. ESA later changed their mission to the European Jupiter probe known as Laplace and costing $800 million. Neither of these ideas were funded, but NASA continues a $10 million annual budget to keep its Jupiter

Europa mission planning alive. The Jupiter Europa spacecraft would be able to orbit Europa and build global maps of the moon's surface, topography and composition. A ground-penetrating radar and gravity-measuring sensors would also probe Europa's interior to obtain definitive proof whether the underground ocean exists. Europe's Jovian probe would mirror NASA's in-depth scrutiny of Europa at Ganymede, which is the largest of Jupiter's moons, as well as the largest natural satellite in the solar system.

NASA's Titan/Saturn System mission would include a NASA-built orbiter to study Saturn and its moons, as well as European lander and research balloon to continue the exploration of the planet's cloud-covered moon Titan. Saturn's moon Enceladus, which harbors ice-spewing geysers, is also major target for that mission. ESA officials have referred to their mission to Saturn and Titan as Tandem. The Titan Saturn System mission would take about 10 1/2 years to reach Saturn if it launched in the 2020 timeframe. NASAs orbiter would spend about two years circling Saturn to study the planet, Enceladus and other moons, and then spend about 1 1/2 years in orbit around Titan.

There are also missions under planning for Uranus and Neptune. Uranus Pathfinder, was recently considered by the European Space Agency (ESA), but was ultimately not funded by ESA. The best time to launch a mission to Uranus would be sometime in the early 2020s, when the planet alignment would be prime for such a long journey. Depending on the chosen propulsion technology, the trip would take roughly 10 to 15 years. The ballpark estimate for a Uranus mission was between $1.5 billion and $2.7 billion.

In March 2011, the National Research Council released its vision for Planetary Science in the coming decade 2013-2022. This Decadal Survey ranked the priorities for flagship missions and gave the two top spots to a Mars sample return mission (MAX-C) and a Jupiter Europa Orbiter (JEO). For the third highest priority, the

Interplanetary Travel

Survey chose the "Uranus Orbiter with Probe", but they said it was unlikely that NASA can execute the Survey's recommendation to initiate the Uranus Orbiter with Probe mission along with MAX-C and JEO in this decade.

A Neptune Orbiter was also proposed by NASA and envisioned to be launched sometime around 2016 and take 8 to 12 years to reach the planet. The California Institute of Technology had proposed one mission plan in 2004, while the University of Idaho and Boeing proposed an alternative approach in 2005, but neither appear on NASA's official pages of missions being considered for the future.

In times of budgetary uncertainty, NASA is forced to proceed with only the most reliable mission proposals. This means a lot of thrilling plans to explore other worlds fall by the wayside. The most notable of these, perhaps, was the Titan Mare Explorer. TiME, as it was called, was a low-cost mission proposal in 2009 to send a spacecraft to Titan. The spacecraft was also a boat, and would have splashed down onto one of Titan's lakes.

The European Space Agency has vowed to carry on with its side of the deal, and has since reorganized its Ganymede mission as the Jupiter Icy Moon Explorer (JUICE) set to launch in 2022 and arrive at Jupiter in 2030, which will examine Ganymede's magnetic field as well as its topography, oceans, and atmosphere.

Europa Mission (Credit NASA)

NASA however remains committed to the Jupiter Europa mission. Not only has Congress approved funding for it in a serious way, but NASA announced in 2015 the final selection of the experiments it will carry based on a formal competition process among dozens of scientific teams.

The White House's fiscal year 2016 budget request for NASA, which was released Monday (Feb. 2, 2015), allocates $18.5 billion to the space agency, including $30 million to formulate a mission to Europa. The new budget proposal signals a commitment from the White House that wasn't there before. *"For the first time, the budget supports the formulation and development of a Europa Mission, allowing NASA to begin project formulation, Phase A,"* NASA officials wrote in a summary of the proposed budget. NASA appears to be zeroing in on a mission that would feature multiple flybys of Europa — perhaps something along the lines of the Europa Clipper, a concept being developed by agency scientists and engineers. Once in orbit around Jupiter, the Clipper would make 45 flybys of the 1,900-mile-wide (3,100 kilometers) Europa over the course of 3.5 years, as the concept is currently envisioned. (Image credit: NASA)

There is even new life in a mission to Saturn's moon Titan to study the Kraken Mare, the biggest sea on what is Saturn's biggest moon. In order to study the sea, it will use a submarine. Once the sub was sent to Titan to begin its adventure, it would have to traverse waters that get down to minus 290 degrees Fahrenheit. The sub is expected to be able to go a maximum speed of a little over two miles per hour and use an engine that gets its power from a radio thermal Stirling generator. The mission would take about three months to study the sea area on Titan. A Titan submarine concept has already been designed by NASA Glenn's COMPASS Team and researchers from Applied Research Lab.

Excerpts from NASA's FY16 budget and predictions to FY20

The OSIRIS-REX spacecraft will launch in 2016, travel to a near-Earth asteroid in 2018, and be the first US mission to carry samples from an asteroid back to Earth. The request also supports a potential future mission to Jupiter's moon Europa, providing double the amount of funding as last year's request to proceed with pre-formulation activities…Informed by the results of FY 2014 testing of solar array and thruster designs, Space Technology continues development of a high-powered solar electric propulsion system that will enable orbit transfer and accommodate increasing power demands for satellites, and power the robotic segment of the asteroid redirect mission…NASA continues to support over 55

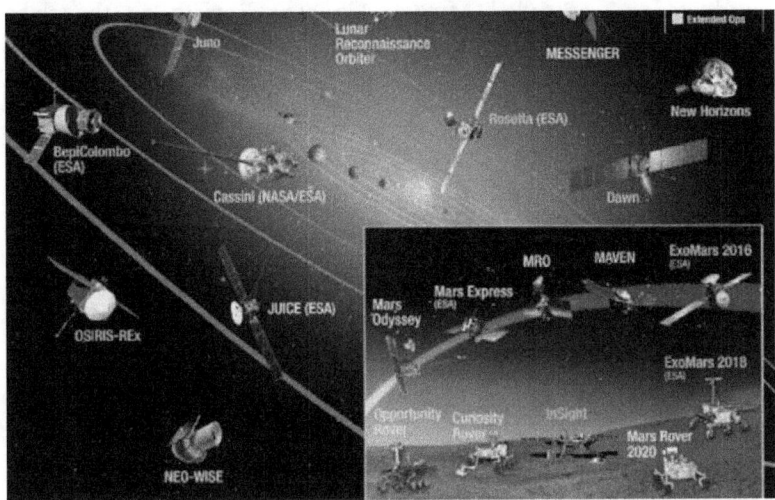

operating Science missions, involving more than 70 spacecraft, many in collaboration with international partners or other US agencies.

The OSIRIS-REx spacecraft will travel to (101955) Bennu, a near-Earth carbonaceous asteroid formerly designated 1999 RQ36, study the asteroid in detail, and bring back a sample (at least 60 grams or 2.1 ounces) to Earth. This sample will yield insight into planet formation and the origin of life, and the data collected at the asteroid will aid in understanding asteroids that can collide with Earth. Launch October 2016.

NASA continued pre-formulation activities for a Europa mission. Most notably, NASA study teams further developed the Europa fly-by mission concept, determining that NASA could accomplish over 80 percent of the science that a Europa Orbiter would achieve for about 50 percent of the cost with a mission that stays in Jupiter orbit and conducts many focused flybys of Europa. NASA completed a series of trade studies, technology efforts, and independent reviews, including determining the technical feasibility of conducting the flyby mission concept with solar power. NASA conducted a Center-led Mission Concept Review on the flyby mission design. NASA also addressed a long-standing risk for a Europa mission by continuing the funding of 15 grants for instrument development and risk reduction under the Instrument Concepts for Europa Exploration program, and solicited proposals for flight instruments for a potential mission to Europa.

Jupiter Icy Moons Explorer (JUICE) instruments development will continue based on the approved schedule. ESA mission adoption, a critical step in the approval of the mission, occurred in November 2014.

Europa instrument proposals will undergo thorough evaluations of science, cost, technical, and management with awards expected in the spring of 2015. Pre-formulation will continue with risk reduction activities in all key science and engineering areas. NASA plans to conduct a Europa mission Agency KDP-A prior to entering formulation in spring 2015. NASA is collaborating with ESA on this ESA-led mission to Ganymede and the Jupiter system. ESA plans to launch the mission in 2022 for arrival at Jupiter in 2030. It has a tentative model payload of 11 scientific instruments. The NASA contribution consists of three separate projects: one full instrument, Ultra Violet Spectrometer; two sensors for the Particle Environment Package suite of instruments; and the transmitter and receiver hardware for the Radar for Icy Moon Exploration instrument.

Jupiter's moon Europa is one of the most likely places to find current life beyond our Earth. For over 15 years NASA has developed concepts to explore Europa and determine if it is habitable based on characteristics of its vast oceans (twice the size of all of Earth's oceans combined), the ice surface – ocean interface, the chemical composition of the intriguing, irregular brown surface areas, and the current geologic activity providing energy to the system. After thorough investigation of concept options, the study teams have identified a flyby concept that delivers the most science for the least cost and risk of all the concepts studied. The flyby concept appears to be feasible based on solar power and without requiring any new technology development, despite the harsh radiation environment that the spacecraft will encounter during the flybys.

NASA will establish a Europa project in FY 2015, initiating the formulation phase. In FY 2016, the project will formulate requirements, architecture, planetary protection requirements, risk identification and mitigation plans, cost and schedule range estimates, and payload accommodation for a potential mission to Europa. The leading mission concept may require significant modification depending on what researchers learn in FY 2015 about the existence of active plumes from Europa's south pole and the accommodations requirements in the awarded instrument proposals.

So the proposed FY16 budget continues a strong program but incorporates important small cuts. For the past two years, Congress has added $80 million and $85 million to NASA's proposed budgets to work on a mission to Europa. The proposed FY16 would reduce funding from the FY15 total Europa budget of $100 million to $30 million. The FY16 budget proposal, like the FY15 proposal, proposes to terminate the Mars Opportunity Rover and the Lunar Reconnaissance Orbiter missions, even though their spacecraft remain healthy, for a savings of $26 million. While the FY16 budget gives the Europa mission its New Start, the funding ramp through 2020 is slow. The budget document doesn't say anything about when the mission would launch or its expected total cost. Based on the slow ramp, the launch seems likely to occur in the mid-2020s. If the eventual mission launches on the SLS rockets NASA is currently developing, flight time to Europa would be about two years versus six and a half years if launch on a commercial rocket. While the SLS seems like the obvious choice, this is an expensive system that has yet to complete development and prove itself while the commercial launchers exist today.

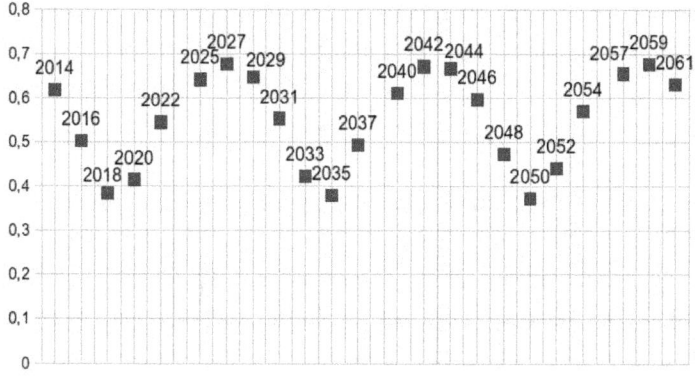

Mars close approaches to Earth, and why 2035 is favored for human flights. Distance from Earth (vertical axis) in AUs. In interplanetary travel the energy needed for transfer between planetary orbits is lowest at intervals fixed by the synodic period. For Earth / Mars trips, this is every 26 months (2 years and 2 months), so missions are typically planned to coincide with one of these launch windows. The energy needed in the low-energy windows varies on roughly a 15-year cycle with the easiest windows (2035) needing only half the energy of the peaks (2027 or 2042).

On December 2, 2014, NASA's Advanced Human Exploration Systems and Operations Mission Director Jason Crusan and Deputy Associate Administrator for Programs James Reuthner announced tentative support for the Boeing "Affordable Mars

Mission Design" including radiation shielding, centrifugal artificial gravity, in-transit consumable resupply, and a lander which can return. Reuthner suggested that if adequate funding was forthcoming, the proposed mission would be expected in the early 2030s.

According to a report in Space News, *"After the initial series of SLS/Orion missions already planned by NASA, the concept would feature crewed "Mars sims" — or simulations — in cislunar space in 2025 and 2027. Those missions would test technologies needed for the Mars missions, including a habitat module and solar electric propulsion. Robotic missions around the same time would test entry, descent and landing technology for future Mars landers. That would be followed in 2033 by the launch of a human mission to Mars orbit. That spacecraft would remain in Mars orbit for a year, possibly landing on either or both of the planet's moons, but not on Mars itself. A crewed "short stay" mission to the surface of Mars would follow in 2039. The architecture does not make explicit use of NASA's Asteroid Redirect Mission (ARM), which the agency has pitched as its next major step towards a human Mars mission. However, workshop co-chairman John Logsdon, director emeritus of George Washington University's Space Policy Institute, said that ARM will demonstrate some technologies, most notably solar electric propulsion, that are needed in this approach."*

More recently, on May 9, 2015 — *"Despite several criticism and doubt, Charles Bolden, the space agency chief said that general agreement is reached upon the mission in NASA's plan to send humans to Mars. "This plan is clear. This plan is affordable, and this plan is sustainable," Bolden said Tuesday (May 5) at the Humans 2 Mars Summit in Washington. NASA astronaut and Russian cosmonaut experimented in investigating the impact a human life could have in a long duration spaceflight. The experiment took one year to complete. Bolden quoted that NASA is preparing to set humans to the locality of Mars by the year 2030. A three-day summit concluded with the objective of sending humans to Mars. Dennis Tito-who, in 2001, was the first to fly to International Space Station. He is running a project aiming to launch two people to Mars around 2021. Another nonprofit organization, Mars One,*

Interplanetary Travel

is aiming to begin a colony in the Red planet with its first colonizers within 2027.

To continue this mission, NASA needs to update its Mars Design Reference Architecture to redirect the technology in hand. NASA has been busy exploring Mars through numerous robotic missions which includes the Curiosity rover. NASA associate admin, Grunsfeld said that the success rate for these robotic missions will certainly differ, compared to a mission conducted by the humans themselves."

No one wants to commit to any mission ideas beyond 2040, but with most activity focused either on establishing the first Mars base or bases in the 2030s, their maintenance and expansion will be an obvious long term goal. Scientific missions to Europa and Titan and outer solar system destinations will slowly be carried out as well, but there are no plans for humans to cross the asteroid belt this century. (Mars habitats: Image credit: NASA)

Getting Around in the Solar System

At Earth's surface, we live at the bottom of a 'gravity well' on the lowest rung of the local potential energy (PE) ladder. The energy 'down here' can be calculated from the formula

$$PE = -\frac{GM}{r}$$

Where G is Newtons Constant of Gravity (6.674×10^{-11}), M is the mass of Earth (5.97×10^{24} kg) and r is the radius of Earth (6378000 meters). When you calculate PE using these values you get -62.4 million Joules per kilogram of matter. If you want to actually break free of Earth's gravity, you have to provide each kilogram of your payload and rocket with exactly 62.4 million joules of kinetic energy to get it to 'infinity'. Because $KE = 1/2 V^2$ per kilogram of matter, by setting the magnitudes PE=KE you get: 62400000 Joules = 0.5 V^2 and so V = 11.2 km/sec is the 'escape velocity' from Earth. The general formula is just

$$V = \sqrt{\frac{2GM}{R}}$$

179

where R is the planet's radius and M is the planet's mass. What is interesting about this is that, suppose you are already at a distance, r from the planet and no longer on its surface (where r=R)? The escape velocity from _that_ distance is still defined by $V = \sqrt{2GM/r}$.

For example, at geosynchronous orbit where we have many communications satellites, r = 42,164 km and so V = 4.3 km/sec, which is only 1/3 the escape speed from Earth's surface. Why is it so low? Let's look at where we are on the energy ladder! At a distance of 42,164 km from the center of Earth, our PE = -9.4 million joules/kg. When you compare this to the -62.4 megaJoules/kg at Earth's surface, it is clear that we have climbed very far out of Earth's gravity well and are high up on the energy ladder some (-9.4 − (-62.4) =) 53 megaJoules/kg from the bottom of the gravity well, which is our first rung on the ladder. In fact, in a few more rungs and another 9.4 mega Joules/kg of energy we are at the end of Earth's gravity ladder, which is under the control of Earth's gravity force, and we can now move onto the energy ladder of either the moon or other distant bodies in the solar system.

Once you are at a particular rung on the energy ladder, you can go into a circular orbit to 'chill out' for a while. To do this is a bit tricky because your strategy for breaking free of the planet and climbing the energy ladder is to travel radially away from the planet's center, but to go into orbit you now have to move at a fixed radial distance with a particular tangential speed. It is physically more efficient to do both of these at once by launching at first vertically from the surface, and then slowly tilting your trajectory over until it is at a tangent to the orbit you want to reach when your speed reaches the orbit speed.

Circular orbits are the easiest to understand because at their distance, the gravitational and centripetal accelerations are exactly in balance so you get $V^2/r = GM/r^2$, or $V^2 = GM/r$.

180

The International Space Station orbits at about 360 km from the surface of Earth (6738 km from the center) and so its orbital speed is just V = 7.68 km/sec. To reach the ISS, your payload has to reach this specific speed tangent to the ISS orbit once the payload also has reached this 360 km altitude. At this altitude, you are now on the energy ladder with a PE of − 59.1 MegaJoules/kg. But this is only the energy you have if you are sitting at the orbit and not actually moving. Once you add your kinetic energy to this, your actual total energy is a bit different.

Vis Viva!

The Vis Viva Equation tells you just how much total energy per kilogram you have at a particular orbit location as a mixture of your orbital kinetic energy and gravitational potential energy.

$$VV = \frac{1}{2}V^2 - \frac{GM}{r}$$

For example, the ISS in its 90-minute circular orbit with a speed of 7.68 km/sec and a distance of 6738 km will have a total energy of PE=-59.1 MJ/kg and KE= 29.5 MJ/kg so VV = -29.6 MJ/kg.

From the surface of Earth to the orbit of the ISS at 6738 km (360 km above surface), V = 7.68 km/s

Potential energy at Earth's surface = GM/(6378 km) = $6.67e^{-11}(5.9e^{24}$ kg)/(6.378×10^6 m) = − 62.4 Million Joules/kg.

At orbit the potential energy is − GM/(6738km) = -59.1 MJ/kg.

Orbit kinetic is $\frac{1}{2}(7680)^2$=29.5 MJ/kg.

Total orbit energy = -59.1 + 29.5 = − 29.6 MJ/kg.

Kinetic energy $1/2V^2$ = 62.4 − 29.6 MJ/kg, = 32.8 MJ/kg so V = 8.1 km/s is the launch speed.

181

Interplanetary Travel

As the rocket climbs to its intended circular orbit, its engines will provide the difference in potential energies between the orbits 62.4 − 59.1 = 3.3 MJ/kg to get to the 360 km altitude. It will then need an additional kinetic energy of 29.5 MJ/kg to match the speed of the orbit. Its total energy change will be 29.5 + 3.3 = 32.8 MJ/kg. In terms of the speed it will need, this KE represents a launch speed of 8.1 km/s.

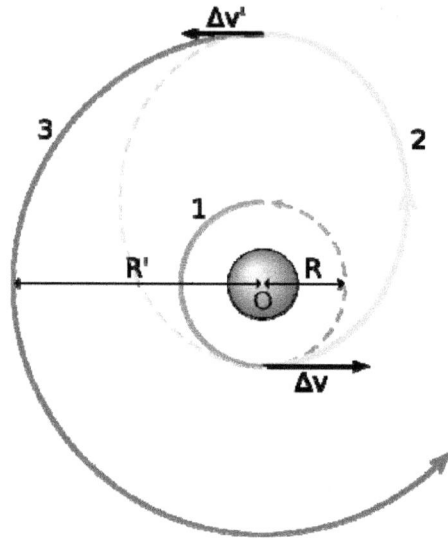

The most conventional way to get from one place to another once you are in space is via a series of Hohmann transfer orbits. The basic idea is that you have two orbits (circular) with distances of R1 and R2 from, say, the Earth, and you want to place your rocket on an elliptical orbit whose perigee is at R1 and whose apogee is at R2, and whose major axis has a length of R1+R2. To do this you have to apply one velocity change at the perigee and a second one at the apogee to get you in and out of the desired circular orbits. Here is a worked example of how this is done:

You want to get a payload from the ISS orbit at 360 km (6738 km orbit perigee) to GEO orbit at 35,786 km (42,164 km orbit apogee):

$GM = (6.67 \times 10^{-11}) \times (5.9 \times 10^{24}) = 3.93 \times 10^{14}$

Orbit Energy $= -GM/(6738+42164) = -8,047,000$ joules/kg

Circular orbit speeds:
Viss=(GM/6738000)$^{1/2}$ = 7642 m/s
Vgeo=(GM/42164000)$^{1/2}$ = 3055 m/s

Vis Viva speeds:
Vtiss = (2 (Viss^2 -8047000))$^{1/2}$ = 10,035 m/s
Vtgeo = (2(Vgeo^2 − 8047000))$^{1/2}$ = 1,603 m/s

Required delta-Vs:
Del-V(T1) = |Vtiss-Viss| = 10,035 − 7642 = 2,393 m/s
Del-V(T2) = |Vtgeo-Vgeo| = 1,603 − 3,055 = 1,452 m/s

So in the ISS orbit, you add a speed of 2,393 m/s to enter the perigee of the Hohmann Transfer ellipse, which has an energy of 8,047,000 joules/kg, then when you reach the apogee of the ellipse at the GEO orbit, you subtract 1,452 m/sec of speed to enter the circular orbit at GEO. The total delta-V your rocket has to provide is 2,393 + 1,452 = 3,845 m/sec. A number of sources have calculated the delta-Vs needed in Hohmann Transfer Orbits to reach many interesting objects in the solar system. On the next page is one provided by Ulysse Carlion based upon calculations using Hohmann transfer orbits and the vis viva equation.

Here's how you read this diagram. Suppose you are in LEO. To get to lunar orbit, you need a delta-V of 3260 m/sec to get captured by the moon, and an additional 680 m/sec to reach Low Lunar Orbit for a total delta-V of 3,940 m/sec from LEO. If you want to go from LEO to Phobos, the delta-Vs will be 3210 + 1060 + 1280 + 3 or a total delta-V from LEO of 5,553 m/sec. If you go from LEO to the lunar surface, however, the total delta-V is 5,670 m/sec which make the lunar surface slightly 'farther' from LEO than Phobos, however the Phobos trip will take eight months and not the mere three days to the moon! That time difference is not an issue for the rocket technology but only one for the human travelers!

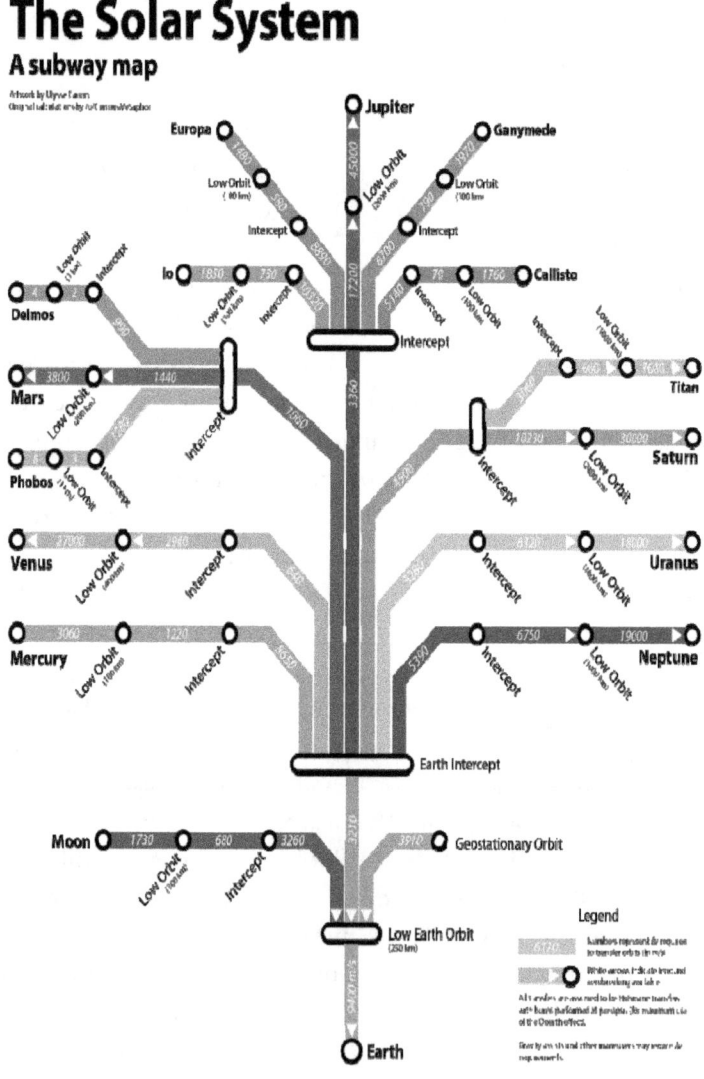

In addition to the familiar low, medium and geosynchronous orbits around a planet (LEO, MEO, GEO) we have several other destinations in space called the Lagrangian points.

L1 and L2 are located on the line connecting the centers of the Sun and Earth. Because the mass of Earth is much smaller than the sun, the distance from Earth to L1 and L2 are equal and defined by $r = D(\text{Mass of Earth} / 3 \times \text{Mass of Sun})^{1/3} = 1.5$ million km, where $D = 149$ million km, Mearth$=5.9 \times 10^{24}$ kg and Msun$=2 \times 10^{30}$ kg.

L4 and L5 are equidistant from Earth along Earth's orbit and form an equilateral triangle between the sun and Earth. Finally, L3 is on the far side of the sun on Earth's orbit and is unstable because of the gravitational influences of the other planets.

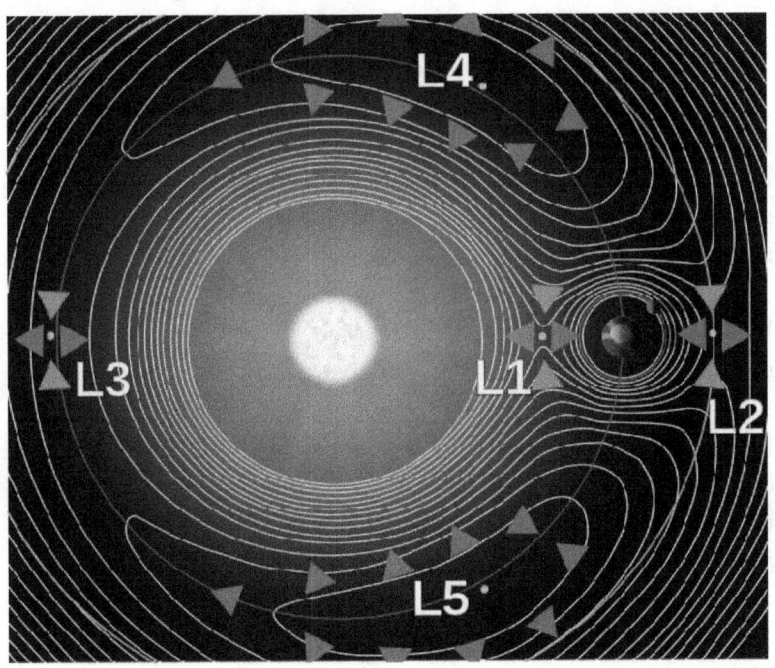

Interplanetary Travel

L1 and L2 have become popular locations for research spacecraft. Including ISEE-3, ACE, Discovr, SOHO and WIND located at L1, and WMAP, Herschel, Planck, Chang'e 2, and JWST at L2.

Across the solar system, there are many objects that occupy the L4 and L5 'trojan' positions. In the case of Jupiter, these 'equilateral points' are occupied by families of Trojan asteroids The Jupiter Trojans, are a large group of objects that share the orbit of the planet Jupiter around the Sun. Mars has 7 asteroids at these locations. Earth has one asteroid 2010 TK7 at the L4 position. Venus has one asteroid 2013 ND15 at the Venus L4 position. Although the Saturn Trojan positions are unstable due to the presence of Jupiter, Uranus has one Trojan asteroid 2011 QF99 at L4, and Neptune has nine Trojans at L4 and three at L5. The Neptunian Trojans are 150 km in diameter and are probably captured Kuiper Belt objects. Pluto is unlikely to have Trojans because its orbit takes it close to Neptune, which would strip these Trojans away from Pluto after a few thousand years.

So, there are many L4 and L5 locations across the solar system where colonies could be set up, however in most cases these are already occupied by large asteroids a kilometer or more in diameter, making for a significant collision hazard.

In addition to Lagrange points involving the planet and the sun, there are also Lagrange points involving a planet with its moons. This leads to a very complex network of points in space, especially for the giant planets, which have dozens of moons each. The difficulty, however, is that due to the proximity of the moons to their planet, most of the Lagrange points such as L4 and L5 become unstable very quickly.

The Earth-Moon Lagrange points are the easiest to understand and because there are no other moons, the most stable.

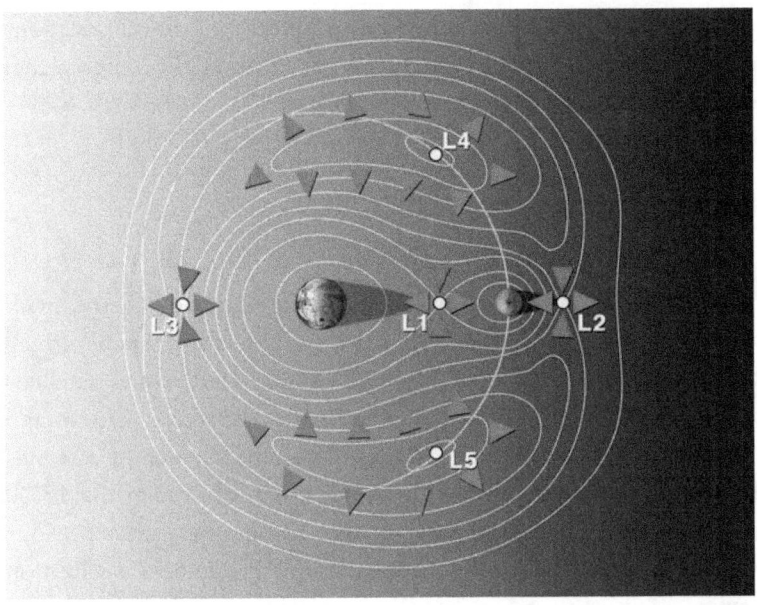

NASA is evaluating an early mission with the Orion capsule placed at Earth-moon L2 (EML-2). Astronauts parked there could teleoperate robots on the lunar farside. EML-2 could serve as a gateway for exploration near-lunar space, asteroids, the moon, the moons of Mars and, ultimately, Mars itself. Meanwhile, the Earth-moon 'trojan' positions (EML-4 and 5) were proposed by Gerard O'Neill for large Earth colonies in 1974, and subsequently advocated by the L5 Society.

At the end of the THEMIS A mission, the spacecraft was renamed ARTEMIS and sent into a halo orbit about the Earth-Moon EML-1 and EML-2. In doing so, ARTEMIS became the first spacecraft to enter halo orbits about the Earth-Moon Lagrange points. The GRAIL mission that recently entered orbit around the Moon performed a low energy ballistic lunar trajectory that flew past the EML-1 point before entering orbit at the Moon. For the most part however, Lagrange points have not been used for missions to or near the Moon.

187

Interplanetary Travel

Just as we have considered a number of interesting travel locations in the Earth-moon-Mars system, we can look at the outer planet satellite systems in the same way.

The Jupiter System

The four Galilean large moon of Jupiter are shown together with the four inner moons: Metis, Adrastea, Amalthea and Thebe, which orbit within 221,000 km of Jupiter's center. The labeled satellites are the main Galilean satellites, which orbit between 421,000 and 1.88 million kilometers from Jupiter. The other 59 minor satellites of Jupiter are shown here, and form a complex system out to 30 million kilometers, and are smaller than 170 km in diameter. Most, in fact, are less than 10 km. (Image NASA/Science Visualization Service)

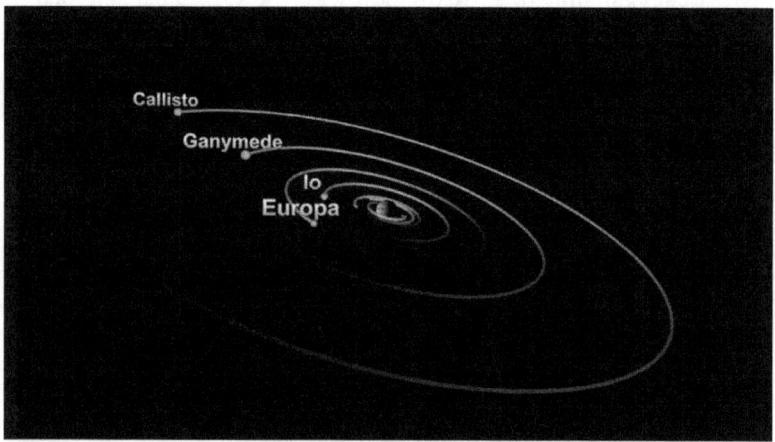

Unlike Mars whose orbital environment is quite calm, Jupiter has a massive magnetic field and radiation belt system. Jupiter would deliver about 36 Sv (3600 rem) per day to Io and about 5.4 Sv (540 rems) per day at Europa. Ganymede receives about 0.08 Sv (8 rem) of radiation per day, and Callisto receives about 0.0001 Sv (0.01

rem) a day. For comparison, the average amount of radiation on Earth is about 0.035 Sv (350 milliRem) per year. Note that 0.75 Sv (75 rems) over a period of a few days is enough to cause radiation poisoning, and about 5 Sv over a few days is fatal. Callisto is, therefore, the only Galilean satellite for which human exploration is feasible without considerable amounts of shielding.

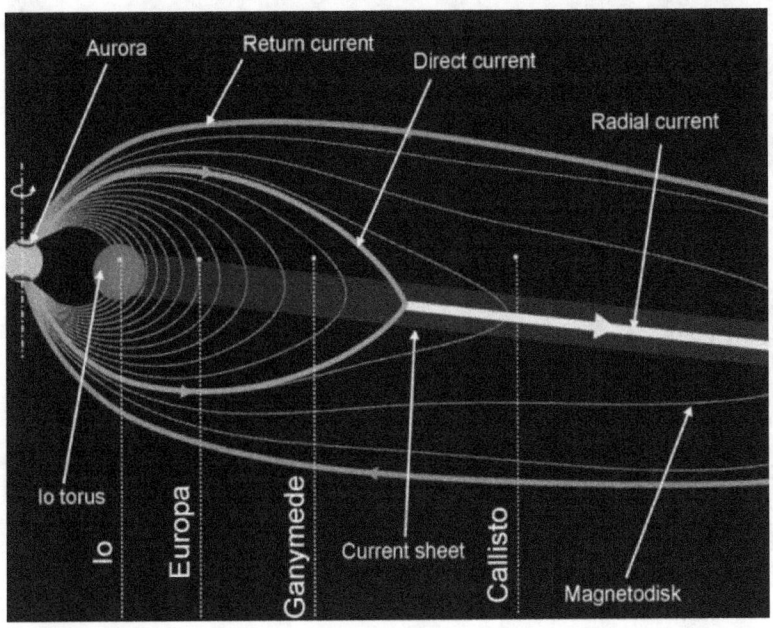

The abundance of water on Europa is a benefit to any considerations for colonization. The colonization of Europa presents numerous difficulties. As Europa receives 540 rem of radiation per day (500 rem is a fatal dose) A human would not survive at or near the surface of Europa for long without significant radiation shielding. Another problem is that the surface temperature of Europa is −170 °C (103 K) (-275°F). However, the fact that liquid water is believed to exist below Europa's icy surface, along with the fact that colonists would spend much of their time under the ice sheet in order to shield themselves from radiation, may somewhat mitigate the problems associated with low surface temperatures.

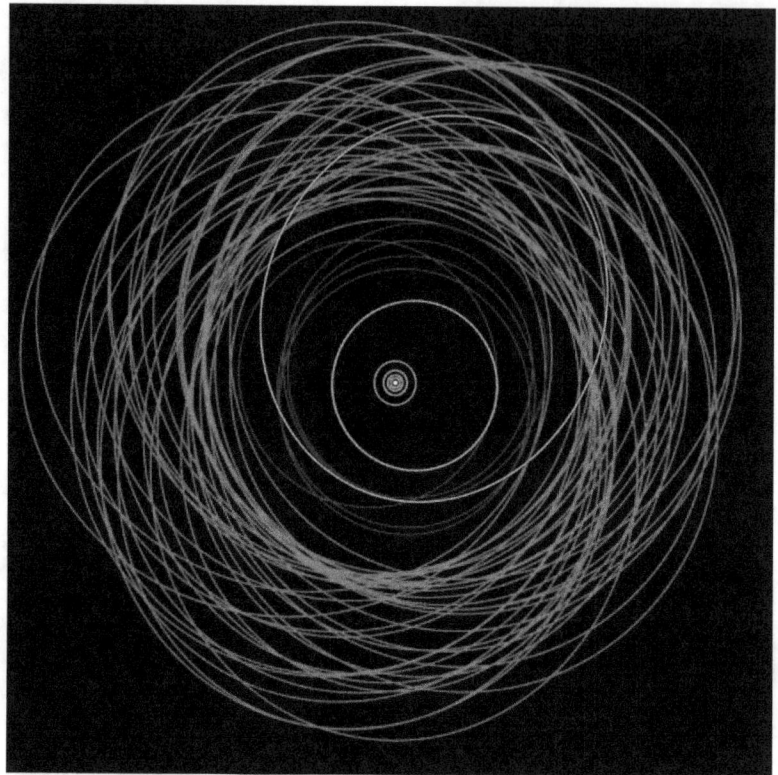

(Image credit: Scott Sheppard)

In Heinlein's *Farmer in the Sky* pioneers on Jupiter's moon Ganymede created their "soil" from scratch by pulverizing boulders and lava flows. Tough work, but in fact it will be even harder for any future farmers there. Heinlein didn't worry about radiation, but we now know he should have. Ganymede receives about 0.08 Sv (8 rem) of radiation per day. Adequate protective measures are hard, short of burrowing deep under the ice and rock.

Using our "delta-V' diagram to put the energy expenditures in perspective, here are the Hohmann pathways to get to the Jovian system from Earth, and some of its larger moons.

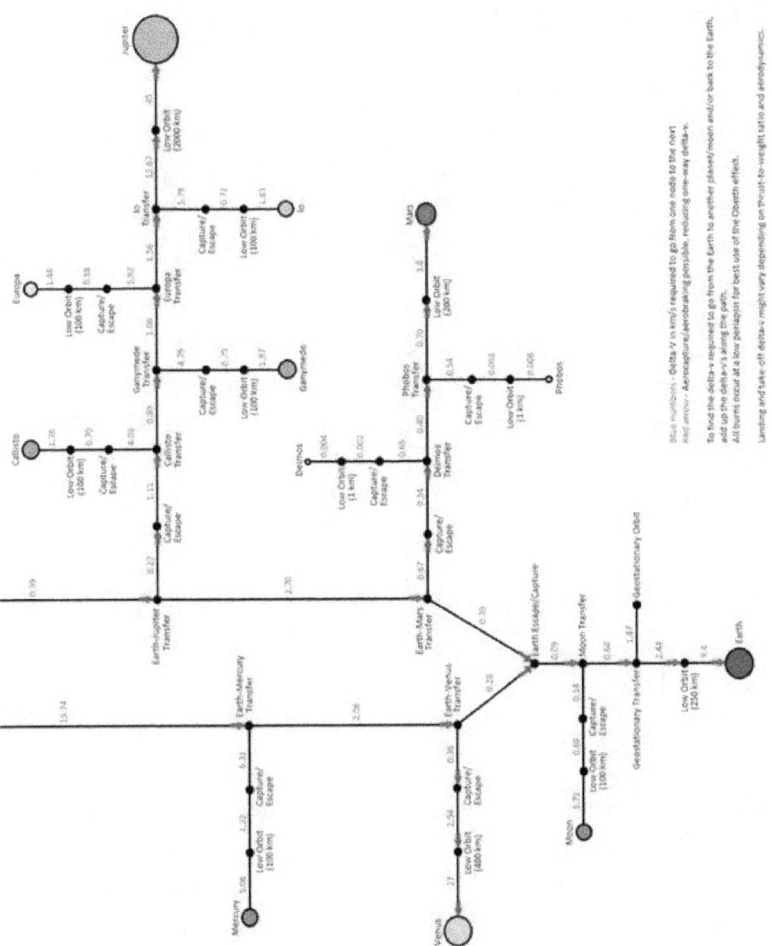

From LEO at Earth, the total delta-V to a 100-km orbit around Callisto is 12.4 km/sec. For Ganymede it is 14.1 km/s, for Europa it is 16.2 km/s and for Io it is 17.7 km/s from Earth LEO.

The Saturn System

A mere 1 km/s further out than Jupiter we encounter Saturn. Because of the sun's weakening gravity, the energy needed to journey from the orbit of Jupiter to Saturn is less than we need to travel from LEO to GEO at Earth! Once captured by Saturn's gravity, we have a number of options of where to go.

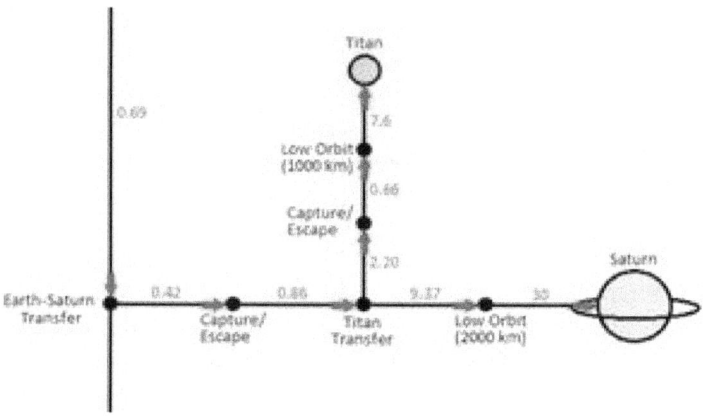

The Saturn system can be thought of as two separate regions of space. The inner region is dominated by the ring system and the moons that have a physical connection with ring structure through complex orbital resonances. This region includes 21 moons out to the orbit of Rhea. (Image: Wikipedia/NASA)

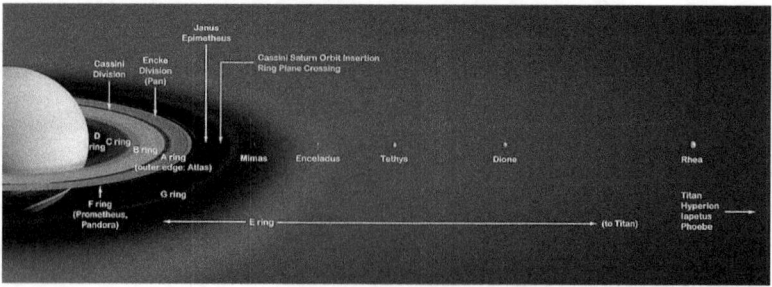

The largest of these moons are Mimas (396 km), Enceladus (504 km), Tethys (1062 km), Dione (1123 km) and Rhea (1527 km). The surface area of Mimas is slightly less than the land area of Spain. The low density of Mimas, 1.15 g/cm³, indicates that it is composed mostly of water ice with only a small amount of rock. Enceladus mass estimates from the Voyager program missions suggested that it was composed almost entirely of water ice. However, based on the effects of Enceladus's gravity on Cassini, its mass was determined to be much higher than previously thought, yielding a density of 1.61 g/cm³. This density is higher than Saturn's other mid-sized icy satellites, indicating that Enceladus contains a greater percentage of silicates and iron. It is an active world with geysers that eject water. Tethys has a low density of 0.98 g/cm3, the lowest of all the major moons in the Solar System, indicating that it is made of water ice with just a small fraction of rock. This is confirmed by the spectroscopy of its surface, which identified water ice as the dominant surface material. Dione is composed primarily of water ice, but as the third densest of Saturn's moons (after Enceladus and Titan, whose density is increased by gravitational compression) it must have a considerable fraction (~46%) of denser material like silicate rock in its interior. Rhea is an icy body with a density of about 1.236 g/cm³. This low density indicates that it is made of ~25% rock (density ~3.25 g/cm³) and ~75% water ice (density ~0.93 g/cm³).

Although Rhea is the ninth-largest moon, it is only the tenth-most-massive moon. Earlier it was assumed that Rhea had a rocky core in the center. However measurements taken during a close flyby by the Cassini orbiter in 2005 cast this into doubt, and the interior is believed to be a homogenous mixture of rock and ice.

Basically, these large inner moons out to an orbital distance of a half-million kilometers from Saturn are giant balls of ice with no detectable surface rocks suitable for mining.

The next group of Saturn's 67 moons include the last of the very large satellites: Titan (5151 km), Hyperion (270 km), Iapetus (1468 km) and Phoebe (213 km). Like most of Saturn's moons, Hyperion's low density (0.544 gm/cm³) indicates that it is composed largely of water ice with only a small amount of rock. It is thought that Hyperion may be similar to a loosely accreted pile of rubble in its physical composition. The low density of Iapetus (1.09 gm/cm³) indicates that it is mostly composed of ice, with only a small (~20%) amount of rocky materials. Phoebe, with a

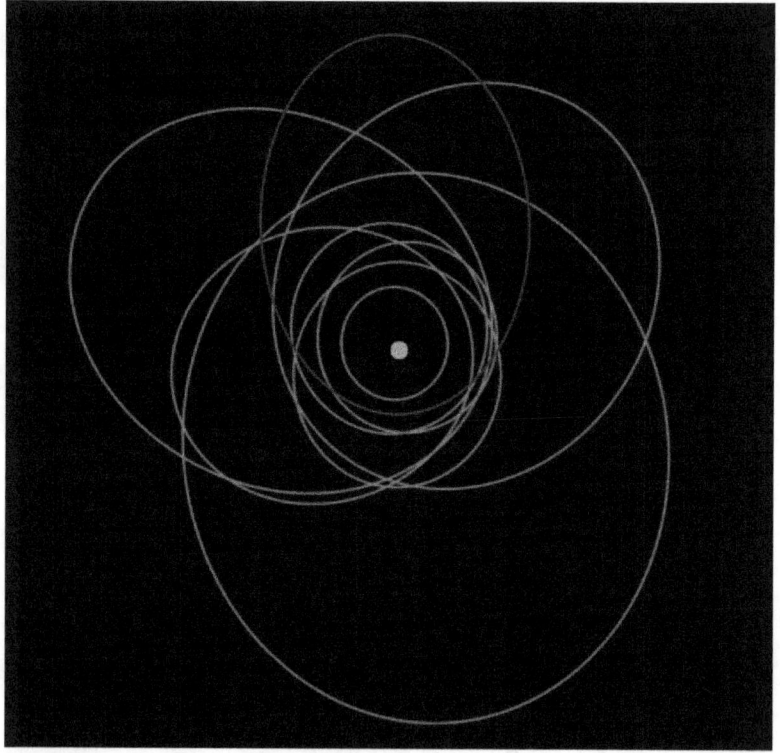

(Image credit: Scott Sheppard)

density of 1.36 gm/cm³, is about 50% rock, as opposed to the 35% or so that typifies Saturn's inner moons. For these reasons, scientists are coming to believe that Phoebe is in fact a captured

Centaur, one of a number of icy planetoids from the Kuiper belt that orbit the Sun between Jupiter and Neptune.

The bottom line is that, since the satellites and ring system of Saturn are predominantly made from ice and very little rocky material is exposed on accessible surfaces for mining, this system will be expensive to explore in terms of delta-V and energy for very little commercial return. <u>Far less expensive</u> ice can be harvested from the outer moons of Jupiter with nearly identical compositions. The major difference being that some of the outer moons of Jupiter are likely to be captured rocky asteroids that could be mined for resources to set up colonies in the Jovian system. No such opportunity exists for Saturn so all raw materials will have to be brought to Saturn from the inner solar system.

Titan is the only well-known exceptional satellite of Saturn that shows enormous promise, though is not without severe challenges. Titan's diameter is 50% larger than Earth's natural satellite, the Moon, and it is 80% more massive. It is the second-largest moon in the Solar System, after Jupiter's moon Ganymede, and is larger by volume than the smallest planet, Mercury, although only 40% as massive. Based on its bulk density of 1.88 g/cm³, Titan's bulk composition is half water ice and half rocky material. Titan is likely differentiated into several layers with a 3,400-kilometre (2,100 mi) rocky center surrounded by several layers composed of different crystal forms of ice. Its interior may still be hot and there may be a liquid layer consisting of a "magma" composed of water and ammonia between the ice crust and deeper ice layers made of high-pressure forms of ice. Titan is the only known moon with a significant atmosphere, the only nitrogen-rich dense atmosphere in the Solar System aside from Earth's. Titan's atmospheric composition in the stratosphere is 98.4% nitrogen with the remaining 1.6% composed mostly of methane (1.4%) and hydrogen (0.1–0.2%). There are trace amounts of other hydrocarbons, such as ethane, diacetylene, methylacetylene,

acetylene and propane, and of other gases, such as cyanoacetylene, hydrogen cyanide, carbon dioxide, carbon monoxide, cyanogen, argon and helium.

In terms of delta-V, Titan is located 3.7 km/sec from the Saturn-Earth capture point in the system, and to get to its surface requires a further 7.6 km/s velocity change, which can be accomplished by atmospheric aerobraking to land, but will require significant Shuttle-style rocket technology to return to orbit. Nevertheless, one-way scientific studies of its surface have already begun with the Cassini/Huygens lander, which reached its surface in 2005. Due to its high gravity and thick atmosphere, sample return missions will be severely challenged to make the trip back to Earth, but robotic surveys such as those being performed by Curiosity in Mars are likely to be an eventual reality, provided that the challenges of mechanically operating at temperatures of 94 k (-179 C) can be overcome.

The abundance of complex organic molecules in the atmosphere an on the surface of Titan may make some form of resource mining possible, but at considerable cost per kilogram.

The remaining moons of Saturn located between 15 and 24 million kilometers from Saturn are small icy bodies barely 20 km in diameter. Many of these moons are in high inclination orbits making them very expensive to get to in terms of delta-V and fuel.

Saturn, Uranus, and Neptune each have magnetic fields with strengths comparable to Earth's magnetic field strength (dipole moment of 7.91×10^{15} T·m^3). The data in the table below shows that the magnetic field strengths for Saturn, Uranus, and Neptune are indeed comparable to Earth, but also that these fields are somewhat stronger than Earth's magnetic field. Consequently, the planetary radiation belts surrounding these planets are not expected to be much larger than Earth's radiation belts. Thus, these outer planet radiation belts do not pose as great of a threat to spacecraft and astronauts as does Jupiter's radiation belts.

The Uranian System

At a further delta-V of only 0.7 km/s we arrive at the satellite system of Uranus. Uranian moons are divided into three groups: thirteen inner moons, five major moons, and nine irregular moons. The inner moons are small dark bodies that share common properties and origins with the planet's rings. The five major moons are massive enough to have achieved hydrostatic equilibrium, and four of them show signs of internally driven processes such as canyon formation and volcanism on their surfaces. The largest of these five, Titania, is the eighth-largest

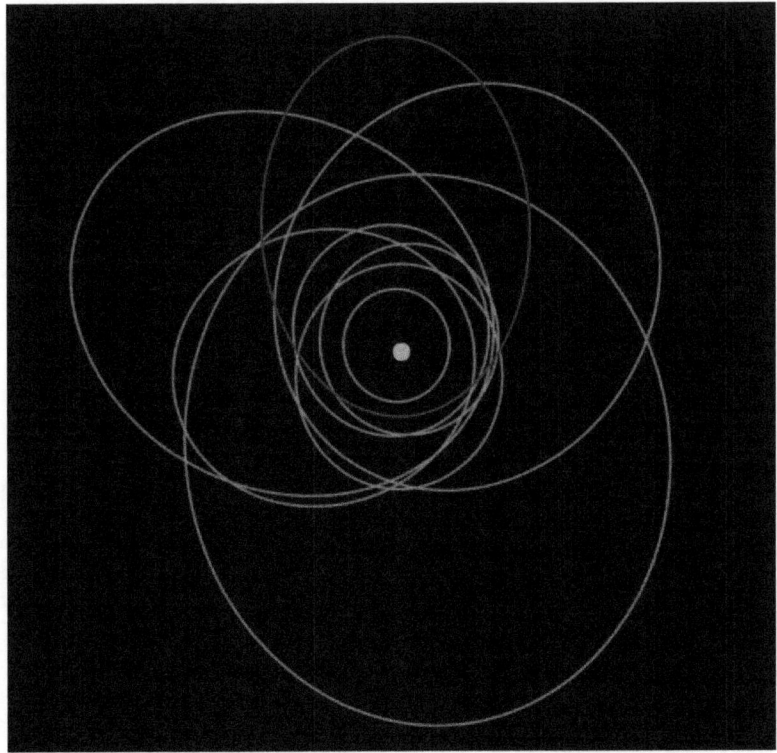

moon in the Solar System, and about one-twentieth the mass of the Moon. (Image credit: Scott Sheppard)

All major moons comprise approximately equal amounts rock and ice, except Miranda, which is made primarily of ice. The ice component may include ammonia and carbon dioxide. Their surfaces are heavily cratered, though all of them (except Umbriel) show signs of endogenic resurfacing in the form of lineaments (canyons) and, in the case of Miranda, ovoid race-track like structures called coronae.

The Uranian system has a unique configuration among those of the planets because its axis of rotation is tilted sideways, nearly into the plane of its revolution about the Sun. Its north and south poles therefore lie where most other planets have their equators. What this means is that we will arrive at a moon of Uranus approaching it from a polar orbit plane rather than the customary equatorial plane. There is nothing bad about this situation compared to equatorial orbits, in fact for survey purposes it is the ideal situation and it costs us nothing extra. Had we entered an equatorial orbit first, we would probably have wanted to place a satellite in a polar orbit to efficiently map the entire surface as it rotates beneath the satellite.

Uranus has a magnetosphere, but it is not filled with the same quantity of energetic particles as Earth's van Allen belts. There is no evidence of helium ions captured from the solar wind, or other ions provided by the surfaces of the moons. Aurora have been spotted near the poles of Uranus, probably created by weak flows of ions from Miranda. The radiation belts are about as strong as those of Saturn.

The Neptune System

Only 0.3 km/sec further out than Uranus is the Neptune system with its 14 moons. The largest moons are Triton (2705 km), Proteus (420 km), Nereid (340 km) and Larissa (194 km). Triton has a surface of mostly frozen nitrogen, a mostly water ice crust, an icy mantle and a substantial core of rock and metal. The core makes up two-thirds of its total mass. Triton has a mean density of 2.061 g/cm^3 and is composed of approximately 15–35% water ice. Triton is one of the few moons in the Solar System known to be geologically active. As a consequence, its surface is relatively young, with a complex geological history revealed in intricate cryovolcanic and tectonic terrains. Part of its crust is dotted with geysers thought to erupt nitrogen. Triton is one of the coldest bodies in the Solar System, with a surface temperature of about 38 K (−235.2 °C). Its surface is covered by nitrogen, methane, carbon dioxide and water ices and has a high geometric albedo of more than 70%. Surface features include the large southern polar cap, older cratered planes cross-cut by graben and scarps, as well as youthful features probably formed by endogenic processes like cryovolcanism. Voyager 2 observations revealed a number of active geysers within the polar cap heated by the Sun, which eject plumes to the height of up to 8 km. Triton has a relatively high density, indicating that rocks constitute about two thirds of its mass, and ices (mainly water ice) the remaining one third. There may be a layer of liquid water deep inside Triton, forming a subterranean ocean.

Chemical reactions caused by radiation have turned the surface on one of Neptune's newly discovered satellites, 1989 N1, as black as chimney soot. The reddish and pink hues of Triton are also the result of such chemical transformations.

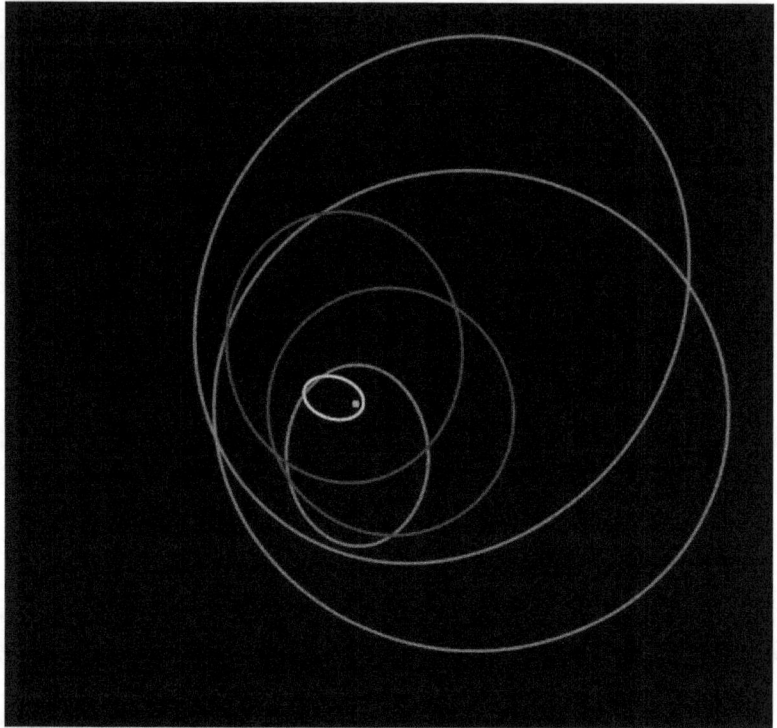

(Image credit: Scott Sheppard)

The prominent features of Neptune's magnetosphere include a radiation belt confined by the orbit of Triton, the compressed field on the dayside, and the long magnetic tail. In the region of the inner satellites of Neptune, the radiation belts have a complicated structure, which provides some constraints on the magnetic field geometry of the inner magnetosphere. Protons have significantly lower fluxes than electrons throughout the magnetosphere, with large anisotropies due to radial intensity gradients. The radiation belts resemble those of Uranus to the extent allowed by the different locations of the satellites, which limit the flux at each planet. Energetic Charged Particles in the Magnetosphere of Neptune

These belts are generally confined inside the orbit of Triton, Neptune's largest moon, in a doughnut-shaped region with Neptune at its center. There is also a cloud of hydrogen surrounding Neptune along Triton's orbit, again in the form of a doughnut. Voyager found that the ionized gas trapped within Neptune's magnetic field consists of two components. One part, mostly protons and nitrogen, has temperatures ranging from about 100,000 to 200,000 degrees Celsius (about 200,000 to 400,000 degrees Fahrenheit); the other, mostly protons, has temperatures of over a billion degrees--the hottest plasma temperatures measured in any of the planetary magnetospheres encountered by Voyager. Since this plasma is exceedingly dilute (a few ions per cubic foot), the heat content is minuscule and therefore did not heat the spacecraft to any measurable extent. Some of this gas finds its way to the upper atmosphere of the planet, producing an aurora, just as is the case for Earth and the other planets with sizable magnetospheres. The aurora on the dark side of Neptune is relatively weak, emitting only about 10 million watts of power, compared to emissions at Earth that range to at least 100 billion watts.

The Asteroids

A reconnaissance of the outer planets and their moons has turned up many disappointments. The major problem is that the accessible resources of Saturn, Uranus and Neptune consist of icy bodies with few rocky silicates that can be mined to fabricate stations and colonies for future human inhabitants. Whatever metals and rocks that will be needed will have to be brought from the inner solar system out to these regions, either as finished habitations or as the raw resources for creating them in situ.

We know that the asteroid belt between the orbits of Mars and Jupiter, along with some of the outer captured satellites of Jupiter, are rich in silicates. Among the outer moons of Jupiter are some that are small but considerably denser than water:

Name	Distance (MegaKm)	Inclin.	Diameter Km	Density g/cc
Ananke	21.3	148	28	2.6
Carme	23.4	164	46	2.6
Pasiphae	24.1	145	40	2.6
Sinope	23.5	128	38	2.6

Ananke has color similar to P-type asteroids, while Pasiphae has color similar to C-type asteroids. Sinope infrared spectrum similar to D-type asteroids.

P-type asteroids have a composition consistent with organic rich silicates, carbon and anhydrous silicates, possibly with water ice in their interior.

D-type asteroids may have a composition of organic-rich silicates, carbon and anhydrous silicates, possibly with water ice in their interiors. They are found in the outer asteroid belt and beyond; examples are 152 Atala, and 944 Hidalgo as well as the majority of Jupiter trojans.

C-type asteroids are carbonaceous asteroids. They are the most common variety, forming around 75% of known asteroids, and an even higher percentage in the outer part of the asteroid belt beyond 2.7 AU, which is dominated by this asteroid type. The proportion of C-types may actually be greater than this, because C-types are much darker than most other asteroid types except D-types and others common only at the extreme outer edge of the asteroid belt. Asteroid 253 Mathilde is an example of a C-type asteroid.

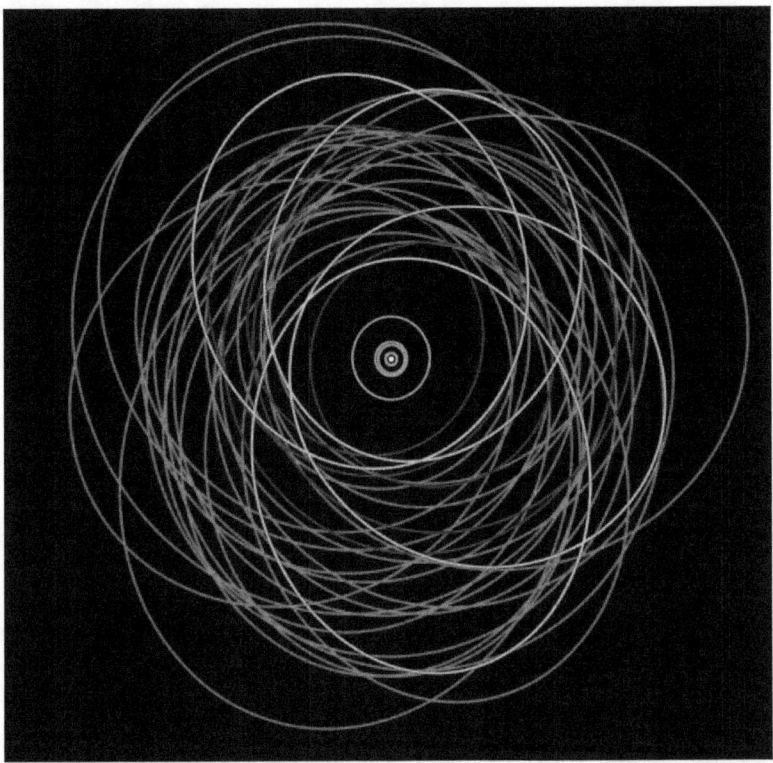

(Image credit: Scott Sheppard)

Saturn's moons all have densities consistent with ice as the major constituent with very little heavier silicates. The distant moon Ymir located at 23 million kilometers is 18 km in diameter, has an estimated mass of 5×10^{15} kg, and a density of 1.7 gm/cm^3 which suggests it is mostly ice but may contain a mixture of silicates and heavier material.

Pluto and Beyond

The dwarf planet Pluto is one of the largest and most famous KBOs.

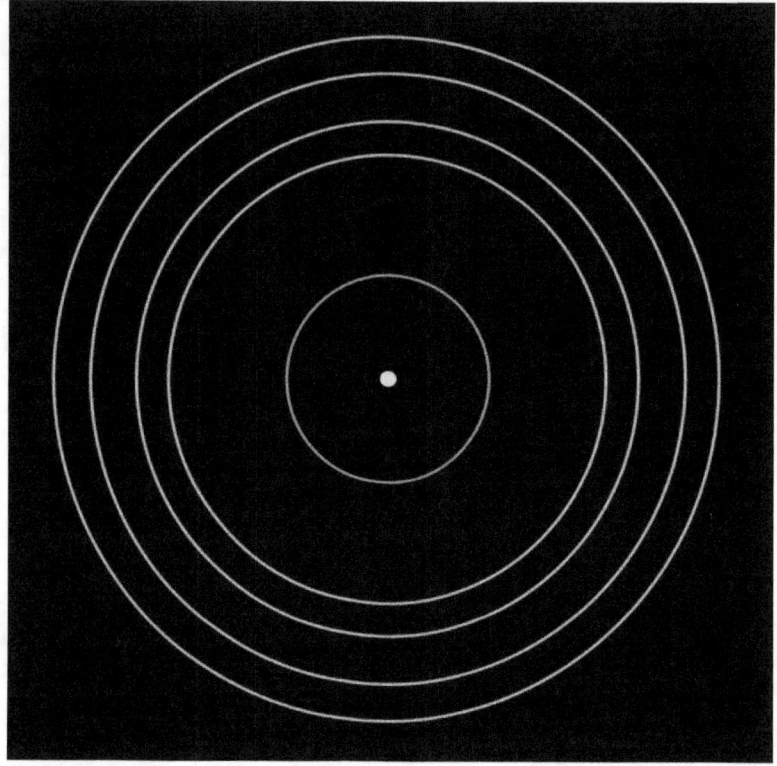

(Image credit: Scott Sheppard)

Its largest moon Charon (1207 km) is in synchronous orbit with a period of 6.4 hours. Charon's volume and mass allow calculation of its density from which it can be determined that Charon is largely an icy body and contains less rock by proportion than Pluto. This supports the idea that Charon was created by a giant impact into Pluto's icy mantle. Observations by the Hubble Space Telescope place Pluto's density at between 1.8 and 2.1 g/cm^3, suggesting its internal composition consists of roughly 50–70 percent rock and 30–50 percent ice by mass. Because the decay of radioactive

elements would eventually heat the ices enough for the rock to separate from them, scientists expect that Pluto's internal structure is differentiated, with the rocky material having settled into a dense core surrounded by a mantle of ice. Pluto's atmosphere consists of a thin envelope of nitrogen (N_2), methane (CH_4), and carbon monoxide (CO) gases, which are derived from the ices of these substances on its surface. Its surface pressure ranges from 6.5 to 24 μbar (0.65 to 2.4 Pa). Pluto's elongated orbit is predicted to have a major effect on its atmosphere: as Pluto moves away from the Sun, its atmosphere should gradually freeze out, and fall to the ground.

Interplanetary Travel

Artist's impression of how the surface of Pluto might look, according to one of the two models that a team of astronomers has developed to account for the observed properties of Pluto's atmosphere, as studied with CRIRES. The image shows patches of pure methane on the surface. At the distance of Pluto, the Sun appears about 1000 times fainter than on Earth. (Image credit: Wikipedia/ESO/L. Calcada)

So, what can we conclude from this story so far? We have surveyed all of the major objects in the solar system and examined their accessibility and mining potential, only to discover some rather startling facts about our solar system. The most accessible destinations are our own moon, Mars, and the near-Earth objects that pass close to our orbit around the sun. Each of these objects has its good and bad points in terms of the resources they can deliver, but we see a rather well-defined pattern based on what we already know about their compositions. No one single body will have all that we need to support human settlements. Some offer metals that can be mined and turned into walls and bulkheads, but do not offer an abundance of nitrogen or hydrocarbons from which to create or grow food. Those that do offer abundant water and hydrocarbons, likewise, do not offer minable rocky or metallic materials. Mars offers many of the minimum components we need including water, accessible iron ores, and nitrogen, which only require mining and reformatting into useful components (fertilizer drinking water, building material. The overriding question, however, is why would we want to travel to the moons of Uranus and Neptune to mine the same kinds of ices we can get from Jupiter's moons or the asteroid belt? The moons of Uranus and Neptune, moreover, are virtually unminable for acquiring rocky or metallic resources, and are only redundantly good for water and rocket fuel.

Science fiction once promised us opportunities to walk in the tropical forests of Venus, or to ice skate on the canals of Mars with only a very light oxygen mask (Heinlein: *The Red Planet*; and the TV series *Babylon 5*). More realistic science fiction stories of the modern ages (Ben Bova *Mars*) try to communicate the idea that the familiar destinations are not a walk in the park. What they do not communicate so well is the motivation behind wanting to live under a pressurized dome with only spacesuit access to the actual environment outside. In virtually all locations outside the asteroid belt, only ice is readily accessible for mining. What is the nature of the motivation that sees this as a desirable way to live?

Antarctica Amundsen-Scott South Pole Station (Image credit: National Science Foundation/Jeremy Johnson)

Astronaut on ice-bound world (Credit: Interstellar)

Orbital Platforms

The easiest way to visit a planet is to simply go into an ordinary orbit around it. This has been accomplished with great accuracy by many different missions by 2015. So far, the spacecraft that have been placed in planetary orbits are Mercury (MESSENGER), Venus (Magellan, Venus Express), Mars (Mars Odyssey, Mars Express, Mars Reconnaissance Orbiter, Mangalyaan, MAVEN), Jupiter (Galileo, Juno), Saturn (Cassini), Asteroids (Dawn, Hayabusa), Comets (Rosetta). Planets with atmospheres are invaluable because as the spacecraft approaches the planet, it can use an aerobreaking maneuver to slow down its interplanetary speed allowing gravitational capture into an elliptical orbit around the planet. The spacecraft does not have to bring any additional fuel with it to shed the several km/s needed to reach a desired, close-in circular orbit. Although relatively simple to perform for Venus (e.g Magellan) and Mars (e.g. Mars Reconnaissance Orbiter), for Jupiter and Saturn the radiation belts and ring particles make such maneuvers very risky and were not tried for the Galileo and Cassini missions. The upcoming Juno mission to Jupiter will also not use aerobreaking, but as for the other outer planet probes, will simply use rockets to provide the necessary, and large, velocity change.

Selecting a particular moon orbiting another planet is much trickier and there are many considerations that go along with this. Not only do you have to enter into orbit around the planet, but you have to adjust that orbit to be captured by the moon of interest, and then find a reasonably stable orbit around that moon. The size of that orbit will be dictated by how close neighboring moons approach your orbit, and the distance and mass of the planet. Both of these factors will try to gravitationally pull your spacecraft away from the intended moon orbit over time.

There has been considerable interest in the moons of Mars as potential platforms for manned outposts. In terms of delta-V from Earth, they are far more accessible than an actual direct landing on the planet.

Phobos and Diemos

Phobos: Diameter = 27 x 22 x 18 km; Mass = 1×10^{16} kg. Distance from Mars = 9,400 km; Surface gravity = 0.57 cm/sec^2; Density = 1.9 gm/cm^3; Escape speed = 11.4 m/sec

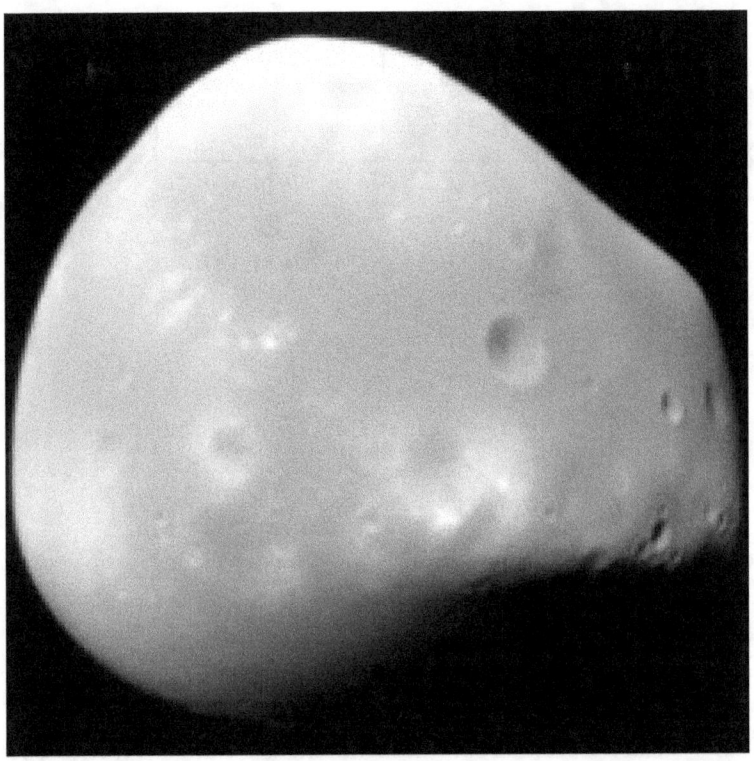

Deimos: Diameter = 15x12 x11 km; Mass = 1.5×10^{15} kg; Distance from Mars = 23,463 km; Surface gravity = 0.3 cm/sec²; Density = 1.5 gm/cm³; Escape speed = 5.6 m/sec

Deimos being farthest from Mars is slightly easier to get to in terms of delta-V by about 300 meters/sec. From Earth LEO, Deimos is at 4.6 km/sec while Phobos is at 4.9 km/sec. Most of this delta-V occurs just leaving LEO and getting to Earth-escape at 3.2 km/sec. As a comparison, landing a spacecraft on the lunar surface requires a delta-V from LEO of 5.67 km/sec, so it is actually easier to get to the moons of Mars than to the surface of our own moon!

Interplanetary Travel

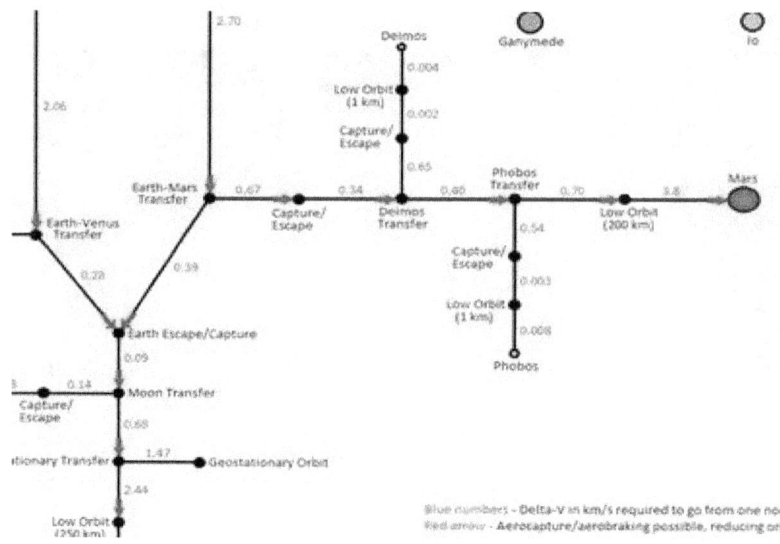

At the distance of Phobos and Deimos, the acceleration of gravity caused by Mars is 47.8 cm/sec² for Phobos and 7.7 cm/sec² for Deimos. If you were attempting to land on either of these moons, the gravity if the moon would be insignificant compared to the gravity acting on you by Mars. In order to avoid being stripped off of the surface of either of these moons by Mars's gravity, you would have to deploy pitons from your spacecraft or lander, and you, yourself would not be able to walk on the day-side of the moons without being ejected into space. Because both moons are close enough to be locked in synchronous rotation with Mars, the day-side would always be the same hemisphere of each moon (similar to our moon as viewed from Earth), and would not be a safe prospecting place without mountaineering equipment and tethers.

Standing up on either of these moons, your horizon would be only $D = (2RH)^{1/2}$, and where your height is $H = 2$ meters the horizon for Phobos would be about 300 meters from where you are standing, and for Deimos it is about 200 meters.

The Soviet Union sent two Phobos missions to deploy landers in ca 1988-1989, and although Phobos 2 returned the first-ever high resolution images of Phobos, the mission failed before the landers could be deployed. In 2011 the Russuan Fobos-Grunt mission to return a sample from Phobos also failed soon after launch.

Meanwhile, scientists and engineers at NASA's Ames Research Center in Moffett Field, Calif., are drawing up a low-cost mission concept that would send a robotic spacecraft to one of Mars' two moons. A robotic mission to Phobos or Deimos could help pave the way for an eventual manned trip to one of these moons. The prospect of a human "base camp" on Phobos or Deimos is enticing to many researchers who envision teleoperating rovers on the Martian surface without the hour-long Earth-Mars time delays, and studying samples launched from the Red Planet up into orbit.

Orbiting the major moons of Jupiter.

Whether you plan to be first-time visitor as an unmanned spacecraft or set up an actual station for hundreds or even thousands of people, you have to find orbits around the jovian moons that are stable. Not only do you have the close passages of adjacent moons to perturb your orbit, but you also have the enormous gravitational force of always-nearby Jupiter to deal with.

One simple calculation for each of the largest moons is, at want orbit distance from each one will the force of gravity of the moon equal that from Jupiter? As your orbits reach this size, they will be more disturbed by the Jovian gravity and eventually pulled away from the moon itself. It is a simple matter to calculate when these gravities will be balanced for every interesting moon as the table shows.

Moon	Diameter (km)	Distance to surface (km)	Acceleration (m/sec2)	Period (minutes)
Io	3660	1040	0.71	33
Europa	3121	1780	0.28	57
Ganymede	5262	6700	0.11	151
Callisto	4820	11,700	0.035	332
Themisto	8	0.5	0.0022	24
Leda	16	12	0.001	74
Himalia	170	600	0.00095	449

Although stable orbits around Themisto and Leda are not likely, Himalia is workable, and at a distance of 11 million kilometers from the terrible radiation belts of Jupiter is probably a very optimum place to park a research station. Although the large Galilean satellites have very Earth-like orbital options, the long term radiation effects are a major problem and shielding will be very expensive unless outposts are literally dug into the crusts of these satellites.

Since all four satellites are in synchronous orbits around Jupiter like our own moon with Earth, ground based outposts on the far side of each moon are tempting, but because of the geometry of the radiation belts, this will not reduce the overall radiation exposure as it rains down the magnetic field lines from 'north to south' on each moon.

Orbiting the major moons of Saturn

Moon	Diameter (km)	Distance to Surface (km)	Acceleration (m/sec2)	Period (Minutes)
Mimas	396	0		
Enceladus	504	0		
Tethys	1062	0		
Dione	1123	0		
Rhea	1527	290	0.136	46
Titan	5151	16200	0.025	454
Hyperion	270	12	0.017	49
Iapetus	1468	5566	0.0029	764
Phoebe	213	1413	0.00023	1413

In the case of the four nearest moons Mimas ,Enceladus, Tethys and Dione, their predicted surface gravities due to their mass is actually less than the gravitational acceleration from Saturn for objects on the moon's surface, so no stable orbits are possible. A similar situation occurs for the more distant moon Hyperion. However, Rhea, Titan, Iapetus and Phoebe appear to have a range of distances where stable orbits may be found.

Orbiting the major moons of Uranus.

Moon	Diameter (km)	Distance to Surface (km)	Acceleration (m/sec2)	Period (Minutes)
Miranda	471	0		
Ariel	1157	172	0.158	36
Umbriel	1169	400	0.081	58
Titania	1576	1950	0.031	156
Oberon	1522	0		
Sycorax	150	0		

Ironically, the most interesting moon, Miranda, with its enigmatic surface features is also one for which a polar observation platform or unmanned spacecraft would not be able to achieve orbit. With Saturn pulling at the satellite with four times Miranda's own gravity, no stable orbits are possible. Ariel, Umbriel and Titania, which look in many ways like 'boring' carbon-copies of our own moon, are the only satellites of Uranus for which a reasonable zone of orbits exist that are stable.

Orbiting the major moons of Neptune.

Moon	Diameter (km)	Distance to Surface (km)	Acceleration (m/sec2)	Period (Minutes)
Larissa	194	0		
Proteus	420	0		
Neired	340	0		
Triton	2705	3750	0.054	161

The significant moons Larissa and Proteus are too close to Neptune for their gravities to overcome the pull of Neptune, and so there are no stable orbits possible for these close-in moons. Neirid is farther away than Triton, but with 1000-times less mass its gravity cannot overcome that of Neptune to provide stable orbits above its surface. Only Triton in this system has the right combination of mass and distance to afford a rather broad 3000-km zone of orbits to offer spacecraft or other platforms.

Orbiting the moons of Pluto

Moon	Diameter (km)	Distance to Surface (km)	Acceleration (m/sec2)	Period (Minutes)
Charon	1207	1125	0.0034	117
Nix	100	46	0.00036	272
Hydra	120	66	0.00021	410

The massive satellite Charon has, by far, the greatest opportunity for a broad range of stable orbits compared to the other low-mass moons of this system. The bizarre and almost mathematical precision with which the orbits of the five moons have been spread out is intriguing from an astrophysical standpoint.

Styx, Nix, Kerberos and Hydra form a 1:3:4:5:6 sequence of near resonances with the Charon–Pluto orbital period, and are spread out about 8,000 km apart within the same orbit plane. (Image credit: NASA/Hubble)

The bottom line from this simple study of the major moons of the planets is that there are remarkably few places where a spacecraft or platform could remain in orbit around a planetary moon large enough to be of interest. A far better option is to find a planet-centered orbit to set up a station or other platform.

It is unclear what the commercial value of these enterprises would be, given that beyond the orbit of Jupiter, the only accessible surfaces are largely ice-bound. They lack sufficient quantities of exposed silicate crust and accessible industrial materials to fabricate habitats or develop a commercial basis for mining activities.

Interplanetary Communication

Nearly 60 years after the start of the Space Age, the number of countries involved in space activities has grown from the two original members, the United States and the Soviet Union, to some 82 space agencies representing nearly as many countries. Not including the multi-national European Space Agency (ESA), there are now ten countries with satellite launch capabilities: USA, Russia, Iran, Israel, North Korea, South Korea, Japan, France, Ukraine, Italy. The vast majority of these efforts involve launching and maintaining communications satellites in Earth orbit, however several of these countries now have embarked on lunar and interplanetary science missions, specifically USA, ESA, Russia, India, Japan and China. USA, ESA, Russia and India have sent spacecraft to Mars, while the USA and ESA remain the only two agencies so far that have sent spacecraft beyond the orbit of Mars into the outer solar system.

To handle the communications challenges of working with spacecraft operating beyond the orbit of the moon, there are now six different 'deep space networks' operated by the USA, Russia, Europe, Japan, China, and India. These networks consist of one or more large radio dishes located at several different geographic locations so that near-continuous reception of extraterrestrial radio signals can be carried out. Although smaller ten-meter-class dishes are sufficient for lunar communication, much larger dishes are generally required to receive more distant signals. The inventory of these larger dishes is not very large. India(32-meter), China (four dishes: 35, 40, 50, 64-meters), ESA (three 35-meter), Japan (64-meter), India (32-meter) and Russia (two 70-meters and one 64-meter). By far the most extensive complex is the United States Deep Space Network consists of three stations (Goldstone, Canberra and Madrid, located about 120 degrees apart in longitude for continuous operations. Goldstone has four 34-meter dishes and a 70-meter dish; Canberra has four 34-meter dishes as well as a 64 and 70-meter dishes. Madrid has three 34-meter dishes and one 70-meter dish. (Canberra DSN station Image credit: NASA)

With these networks, we already have routine communication with a suite of orbiting spacecraft at Mars, Jupiter and Saturn, as well as continued contacts with the New Horizons spacecraft near Pluto.

The Voyager 1 spacecraft now located over 130 AU (20 billion km) from the sun is still in radio contact with Earth through the

Goldstone 70-meter dish. This remarkable image is of the Voyager 1's radio signal detected by the National Radio Astronomy Observatory's Very Long Baseline Array on February 21, 2013. At the time, Voyager 1 was roughly 11.5 billion miles away. The radio image is roughly 0.5 arcseconds across, and the oblong shape of the 22-watt radio signal is a result of the array's  configuration. The 'twinkling' of this radio star at thousands of times per second is what codes the 1s and 0s of the binary signal that transmits data to Earth.(Image credit: NASA/NRAO)

Radio dishes, and state-of-the-art receiver technology at the 70-meter class level, are the minimum buy-in for being able to send and receive radio signals across our entire interplanetary domain out to the Kuiper Belt and the solar heliopause – the nominal boundary of interstellar space at about 140 AU..

There are quite a few factors that enter into successful interplanetary radio communications. Chief among these are the broadcast power, the bandwidth of the communications, and the sizes of the transmission and reception dishes. With a large enough transmission power measured in watts, you can make do with smaller dishes to collect the feeble radio radiation. Also, if you make your bandwidth small enough, you are concentrating this energy across only a small number of parallel frequencies.

The way in which these factors play against each other can be gleaned from the following table, which shows the kinds of systems being used in the current generations of interplanetary spacecraft.

Name	Dish (meter)	Power (watts)	TM rate (bps)	Year	Object
Mariner 10	1.4	10?	118k	1974	Mercury
Mariner 10	1.4	10?	118k	1973	Venus
Magellan	3.7	20?	268k	1990	Venus
Mariner 4	0.1	10	33	1964	Mars
MRO	3.0	100	6000k	2006	Mars
Dawn	1.5	100	41k	2015	Ceres
Voyager 2	3.7	20	115 k	1979	Jupiter
Voyager 2	3.7	20	44k	1981	Saturn
Cassini	4.0	20	160k	2004	Saturn
Voyager 2	3.7	20	21k	1989	Neptune
New Horizons	2.1	12	1000	2015	Pluto
Pioneer 10	2.7	8	16	1997	70 AU
Voyager 2	3.7	20	160	2015	107 AU

Some communications satellites in Earth's geosynchronous orbit such as Intelsat Galaxy 13 have kilowatts of broadcast power with similar dish sizes and transmit UHDTV-quality images at 100 million bits/sec with a 3840×2160-pixel format.

The contrasts between what we are accustomed to in our home entertainment and internet experience, and the engineering realities of current interplanetary communication limited to less than 100-watts of transmission power, are quite striking. The chief problem has to do with the current limitations on the cost of the (non-commercial) payload we can deliver to these different locations across the solar system. It is still largely true that, after nearly 60 years of space activity, it continues to cost about $10,000 per pound ($3,800/kg) to get payloads into Earth orbit. Sending payloads to Mars and beyond is even more expensive. For example, the New Horizons mission which flew by Pluto in 2015 had a mass of 470kg including 80 kg of propellant. It was launched on an Atlas V rocket which cost $170 million, so to Pluto the cost is a whopping $360,000/kg! At these prices, the difference between a 3-meter system at 20 watts and a 10-meter system at a kilowatt broadcast power is spectacular. If we were to send a Galaxy 13 system to Pluto, instead of costing $250 million to GEO, it would have run up a bill to the science budget of NASA of $25 billion.

Scientific payloads have to be shoehorned into the smallest-mass systems that can marginally carry out the research goals and send data back to Earth. That's why scientific missions make due with 3-meter dishes and very low data rates that would barely compare with the old 1980 computer modem technology of 56 kbps. While Earth internet access and HDTV breezes along at tens of megabits/sec, until launch costs come down dramatically, we are stuck with far less than 1 mbps in the inner solar system, and less than 100k bps in the outer solar system.

There are things we could consider doing to improve this situation, but the fault lies entirely with the size of the dish and the transmitting power that we can bring with the spacecraft. When a radio 'dish' transmits a signal, most of the power is radiated through the primary beam of the dish, but the rest about 30% leaks out in other directions through the secondary and tertiary side-lobes of the dish. The ratio of the power transmitted through the primary to the side lobes is called the antenna gain, so for the highest efficiency and least waste of energy, you want a high-gain antenna. This is why dishes are used because more than 70% of the radiated power goes into the main lobe.

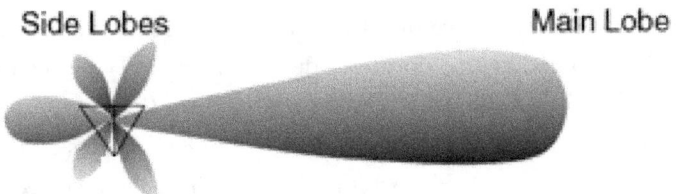

Side Lobes   Main Lobe

$$wavelength = \frac{speed\ of\ light}{frequency}$$

$$= (3x10^{10}\ cm/s) / (3x10^9\ Hz)$$
$$= 10\ centimeters.$$

Since 3 meters = 300 cm:

Theta = 69.9 x (10 cm/300 cm)
= 2.3 degrees

The angular width of the main lobe is given by the formula

$$\theta = 69.9 \frac{\lambda}{D}$$

Operating at a frequency of 3 gigahertz, the wavelength is $\lambda=0.1$ meters, and with a dish diameter of D=3-meters, which is a popular size for current interplanetary spacecraft, you get a primary beam that subtends an angle of $\theta=2.3$ degrees on the sky. That is an angle about 5 times the diameter of the full moon!

$$\frac{2.3\ degrees}{57.3\ degrees} = \frac{X}{7.5x10^9}$$

$$X = 3.0x10^6\ km$$

At the distance of Pluto, which is 7.5 billion km, the width of this beam would be about 300 million kilometers in diameter. That equals the entire diameter of Earth's orbit! That also means whatever energy you put into transmitting the signal from Pluto, 70% of the energy broadcasted by your transmitter will be spread out over a circular area 300 million kilometers ($3x10^{11}$ meters) in diameter!

For every watt of energy you transmit from 1 square meter of your spacecraft dish, only about $0.70/\pi(3x10^{11}/2)^2 = 2.5x10^{-24}$ of the original watts is intercepted by 1 square meter of the receiver on Earth.   To gather this up into a detectable signal, you need a very big radio dish on Earth. For example, the 70-meter Goldstone antenna has an area of $A = \pi (70/2)^2 = 3,846$ meters^2. For a 20-watt signal sent from Pluto with a 3-meter dish, the received power at Goldstone is $P = 20$ watts x $2.5x10^{-24}$ x 3846 = $2x10^{-19}$ watts.

This is a level that is comfortable for Goldstone to detect, but for a dish only 34-meters in diameter, the received signal would be a factor of four weaker ($(70/34)^2 = 4$) and might be undetectable. Even though you could detect this signal at Goldstone, there is another problem to contend with.

If the received signal to Goldstone were just a simple on-off signal lasting one aecond, you could easily detect it if you watched the spacecraft for one second. During that one second, you would be able to add up the weak power from the spacecraft and detect the energy from the on pulse. So your system works real well for a 1 bps signal. But suppose you wanted to transmit two pulses in one second (2 bits/sec). You would have to divide your transmitted power every second by two, so now your signal to Goldstone would be half as strong per pulse! If Goldstone had trouble detecting the 1 bps signal, it might not even be able to detect the 2 bps signal at all! This is why spacecraft have to lower their transmission rates to Earth as they get farther away. This is especially noticeable when spacecraft like Pioneer, Voyager or New Horizon travel beyond the orbit of Saturn. The only way to compensate for the distance is to increase the transmission power to keep the data rate in bps constant. But when spacecraft are only provided a few hundred watts of power to run all of their systems, there is only so much you can do. You can try increasing the diameter of the transmission dish. This reduces the 'beam area' at Earth and concentrates the signal, but at $300,000 per kilogram, you can't build dishes large and light weight enough to make a difference.

Laser-optical communication

One problem with the radio system is that the transmitting dish has such a big beam that the power is distributed over a huge area at the destination. Laser beams, however, are far more intense and can be greatly focused at their much shorter wavelengths. For

225

example, the Apache Point Observatory Lunar Laser-ranging Operation uses a 3.5-meter telescope to transmit pulses of laser light to the moon and to detect their return pulses. The pulse starts out 3.5 meters in diameter as it exits the telescope and is about 2 km wide at the moon.

The angular resolution of this telescope at 1 micron is Theta=69.9 x (10^{-6} meters/3.5 meters) = 0.00002 degrees. At the distance of the moon (384,000 km) this is a circle with a diameter of about 100 meters, but atmospheric effects on Earth broaden this to about 2 kilometers. At a distance of Pluto (7.5×10^9 km), this beam would be about 2,600 km across.

The problem with such a 3.5-meter laser system is that the spacecraft would have to have an identical optical system. Some typical 3.5 meter ground-based telescopes have a mass of about 36 tons, most of which is the mounting. This is too expensive for a spacecraft!

(Image credit: Apache Point Observatory)

How much laser power do you need in order to be seen at the orbit of Pluto? You can buy a commercially available 10 kW 1060-nm-emitting YLS-10000-Y13 fiber laser from *IPG Photonics* (Oxford, MA). These are used for spot-welding! At Pluto, this energy emitted by a 3.5-meter telescope on Earth would be spread out over diameter of 2,600 km, so its brightness would be about 10000 watts/$[\pi \ (2.6 \times 10^6/2)^2 \ ]= \ 2 \times 10^{-9}$ watts/meter^2. A similar, expensive, 3.5-meter telescope on the spacecraft would capture about $\pi \ (3.5/2)^2$ x $2 \times 10^{-9} = 1.9 \times 10^{-8}$ watts of energy. Now at this wavelength, each photon of light carries 2×10^{-19} joules of energy. Since 1 watt = 1 Joule in 1 second, this works out to be about 96 billion photons per second!

Most modern imaging detectors can detect single photons. That means we could take this incoming photon stream from Pluto and break it up into a 100 megabit rate where each bit is defined by 960 photons (100 megabits x 960 photons/bit = 96 billion photons).

(Image credit: Apache Point Observatory)

Interplanetary Travel

That would let us enjoy UHDTV from Pluto just the way that Galaxy 13 provides it from Earth orbit. Because a 3.5 meter optical telescope operating at 1 micron wavelength can have a mass of many tons, we could also consider reducing the mass of the spacecraft optical system enormously until we could actually afford it! But it would also have to have its own laser generator, and those are very massive and expensive as well. The engineering challenge is to balance all these factors so that the spacecraft system is as inexpensive as possible but still provides a massive improvement in data rates across the solar system. NASA engineers are hard at work on this problem and have already made some amazing breakthroughs.

In 2004, NASA created the Mars Laser Communication Demonstration (MLCD) project to demonstrate the first deep space optical link. Lincoln Laboratory/MIT, NASA-JPL and the NASA-Goddard Spaceflight Center were partners in this $300 million effort. The plan was to send a $500 million spacecraft to Mars called the Mars Telecommunications Orbiter with a 5-watts laser system, which would be received by the 200-inch Hale Telescope at Mount Palomar. The expected speeds were in the neighborhood of 10 to 30 million bits/sec compared with the 130,000 bps currently used by Mars missions today. For instance, at the current rate, just one high-resolution image from the Mars Reconnaissance Orbiter currently takes about 1.5 hours to transmit back to Earth. With the proposed laser system, it would only take a few seconds!

MLCD was able to do some of the engineering design of critical aspects of such a system. For example, JPL operated a Mars simulation of such a laser communications system at 60 megabits/s (Mb/s). Receivers for a data rate of 1.6 Gb/s were also under development. The program also established one of the major technological challenges for such an interplanetary system: Pointing.

The angular width of a laser beam transmitted through a 30-cm (11-inch) diameter telescope is approximately 0.7 arcseconds. From Mars, Earth is a disk in the sky with a diameter of about extends about 7 arcseconds. To accurately point the laser at the Earth and avoid loss of signal at the receiver from the distance of Mars, you have to point the laser with an accuracy of 0.35 arcseconds or 1/20 the diameter of Earth. At greater distances from Jupiter, Saturn or farther, the pointing demand increases proportionately. To compensate, you can increase the diameter of the telescope, but that adds mass and greatly increases the cost of the system.

The JPL tests suggested that the full-sized system operating in deep space using a ground-based telescope on Earth would require telescopes as large as 5-10 meters to provide data rates of 1Gb/s from the maximum Mars range, 100 Mb/s from Jupiter distances and 10 Mb/s from Uranus. The spacecraft systems could then be smaller-sized telescopes that meet the mass and cost limits.

A ready-to-go laser communications transceiver. (Image credit: JPL)

Interplanetary Travel

On July 21, 2005, MTO was canceled due to the need to support other short-term goals, including a Hubble servicing mission, Mars Exploration Rover extended mission operations, launch Mars Science Laboratory in 2009, and to prevent the Earth science mission Glory from being cancelled.

Meanwhile, the Lunar Atmosphere and Dust Environment Explorer (LADEE) was launched in 2013 and included a Lunar Laser Communications Demonstrator (LLCD). On October 18, 2013, it successfully transmitted data between the spacecraft and its ground station on Earth at a distance of 385,000 kilometers (239,000 mi). This test set a downlink record of 622 megabits per second (Mbps) from spacecraft to ground, and an "error-free data upload rate of 20 Mbps" from ground station to spacecraft. The previous record from the moon had been 150 megabits per second, achieved by NASA's Lunar Reconnaissance Orbiter (LRO).

"Just imagine the ability to transmit huge amounts of data that would take days in a matter of minutes," LLCD manager Don Cornwell said in a statement. "We believe laser-based communications is the next paradigm shift in future space communications."

In 2011 NASA set up a new program called the Laser Communications Relay Demonstration mission (LCRD), which is slated to lift off in December 2017. By 2012, NASA had invested $3 million to contract with Space Systems/Loral to use a scheduled communications satellite to host LCRD in geosynchronous orbit. The experiment's two optical modules will use lasers to send information to two ground stations, one in California and one in New Mexico, at rates of up to 1.25 gigabytes per second. By 2015, this program was still funded and moving ahead with the building of the flight-ready technology.

Meanwhile, the first gigabit laser-based communication was achieved by the European Space Agency and called the European Data Relay System (EDRS) on November 28, 2014. And on December 9, 2014, NASA's OPALS mission onboard the International Space Station uploaded 175 megabytes in 3.5 seconds. The system was also able to re-acquire tracking after the signal was lost due to cloud cover.

Although work continues on testing laser communications out to lunar distances, there are currently no funded interplanetary missions that plan to use this technology. Nevertheless, laser-based interplanetary communications systems will be the reality of any future activities across our solar system. The severe challenge, however, is that radio communications systems that operate at a snail's pace are already flight tested and established conservative and low-cost technology. It is very hard for new ideas to be adopted in designing flight missions when conservative and seemingly less risky technologies exist. At what point laser-based systems will finally be given their interplanetary opportunity to prove their mettle is very much an open question.

Meanwhile, in the not too distant future, there will exist a vast network some have referred to as the Interplanetary Internet, which will shuttle information across the entire solar system at HDTV rates. The nature of this vast torrent of interplanetary information is anyone's guess. A major need will be to support activities on Mars, and scientific research spacecraft in the outer solar system. The current international focus on Mars will be the largest customer for this technology. As for deep space scientific probes, the scale of the Interplanetary Internet will have to slowly evolve as more of these missions to the outer solar system are funded. Of course, we will never be able to avoid the time delays between transmission and reception. The idea of true 'real-time' communication is a myth.

Radiation Hazards

Radiation comes in two forms: particles and electromagnetic. Particle radiation includes protons, neutrons, electrons and the nuclei of many kinds of atoms such as helium and iron. Electromagnetic radiation includes all forms of light, which travel at 300,000 km/sec and include gamma rays, x-rays, ultraviolet, infrared and radio-forms.

Radiation in all its forms can be measured accurately. We can also, precisely determine at what levels it becomes a human hazard. Scientists measure radiation doses and dose equivalent in terms of units called Rads and Rems (Grays and Seiverts are used in the 'SI' System of meters-kilograms-joules).

Dose: This is a measure of the amount of total energy that is absorbed by matter over a period of time. This matter can be human tissue, or sensitive computer circuitry. The unit for dose is the Rad, which means 'Radiation Absorbed Dose'. One Rad is equal to 100 ergs (10^{-5} joules) of energy delivered to one gram of matter. The equivalent Isp unit is the Gray (G). One Gray equals 100 Rads.

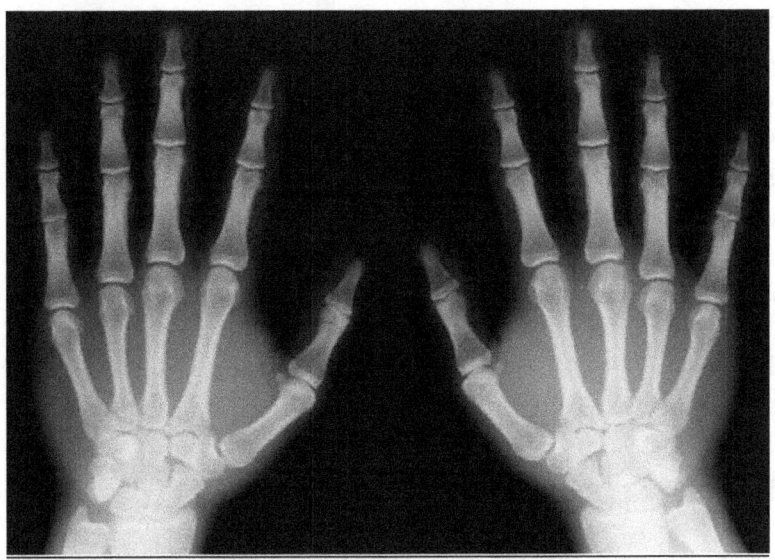

Dose Equivalent: This compares the amount of absorbed energy (Rads) to the amount of tissue damage it produces in a human. It is measured in units of the Rem, which means 'Roentgen Equivalent Man'. The equivalent Isp unit is the Seivert (Sv). One Seivert equals 100 Rems.

Radiation dose is just the amount of energy delivered to a sample of matter. Equivalent dose, however, is much more complicated. This term has to do with the amount of damage that a given amount of energy does to a tissue sample or an electronic component. Each kind of radiation, for the same exposure level, produces a different amount of damage. Mathematically, this is represented by the equation:

Dose Equivalent (in Rem) = Dose (in Rads) x Q
or
Dose Equivalent (in Sieverts) = Dose (in Grays) x Q

Different forms of radiation produce different levels of tissue damage. EM radiation, such as x-rays and gamma-rays, produce 'one unit' of tissue damage, so for this kind of radiation Q = 1, and so 1 Rad = 1 Rem. This is also the case for beta radiation, which has the same Q value. For alpha particles, Q = 15-20, and for neutrons, Q = 10. That means that a dose of 1 Rad of radiation (which equals 100 ergs delivered to 1 gram of matter) produces a dosage of 10 Rem for Q = 10.

Radiation comes from the sun and stars in the form of light, and also in the form of particles due to solar flares and other explosive events. Flares are produced when a star's surface magnetic fields become so twisted they re-configure themselves into simpler shapes. The resulting change liberates more energy than 1000s of hydrogen bombs. Solar or stellar flares are more than a curiosity and a nuisance. We have had many close-calls during the Space Age with intense solar flares that were unpredictable, but fortunately no one was in harm's way.

Dose Equivalent	Health Effect
50 - 100 Rem	No significant illness
100 - 200 Rem	Nausea, vomiting. 10% fatal in 30 days.
200 - 300 Rem	Vomiting. 35% fatal in 30 days.
300 - 400 Rem	Vomiting, diarrhea. 50% fatal in 30 days.
400 - 500 Rem	Hair loss, fever, hemorrhaging in 3wks.
500 - 600 Rem	Internal bleeding. 60% die in 30 days.
1,000 Rem	Intestinal damage. 100% lethal in 14 days.
5,000 Rem	Delirium, Coma: 100% fatal in 7 days.

Radiation comes from the rest of the universe in the form of cosmic rays created by supernova explosions and by certain kinds of galaxies in deep space. Radiation also comes from radioactive minerals such as feldspar (uranium and radon), which is common on the surface of granite-rich planets.

There are also two distinct environments in which travelers will be exposed to these radiations: in space and on a planetary surface.

Planetary Surfaces - On the surface of a planet with an atmosphere, the atmosphere itself provides an enormous natural shield to particles radiation, though even on the surface of Earth we still get a measurable amount of cosmic rays, especially if you live at higher elevations.

The amount for shielding is just the shear mass of atmosphere above your head. For Earth, we have about 240 gm/cm2. For a planet like Mars with 1/10 the atmosphere of Earth, there is hardly any shielding at all. It's about 12 gm/cm2, or about the same as the walls of the International Space Station.

Mars is popularly considered colonizable even though no colonist could ever stroll its surface without the aid of a spacesuit equipped with radiation shielding. On Earth, the cosmic ray exposure is about 26 milliRem or 7% of your <u>annual</u> total at sea-level, and about 50 milliRem or 14% at the elevation of Denver. Typically, our annual radiation dose is about 370 milliRem (or 3.7 milliSieverts). On Mars, the average exposure is about 20 milliRem per <u>day</u>!

Also on the surface of Earth, we have granite-rich rocks in the crust that emit radon gas providing a background dose that we can only minimize by living in areas where there is little of this granite-

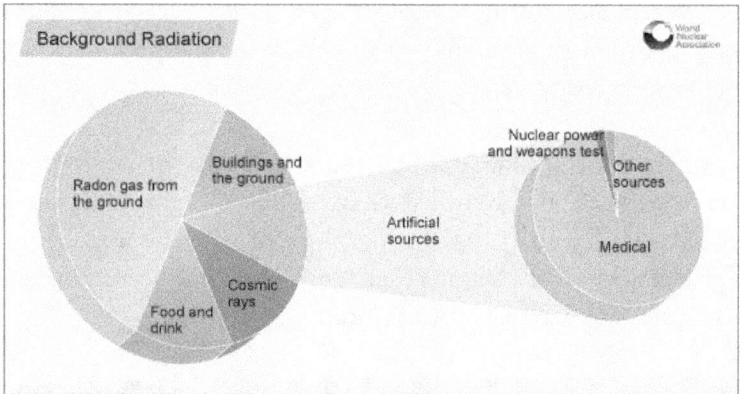

rich material. On Earth, 50% of your annual radiation exposure comes from radon gas generated by clays and building materials! For example, if you decided to live outdoors in a tent, your exposure would be 0%, but if you lived in an unventilated basement recreation room, it would be 50% of your total natural background exposure. Radon exposure is responsible for up to 14% of all lung cancer cases.

Electromagnetic forms of radiation such as x-rays and gamma-rays are shielded from the planetary surface also by virtue of their being a thick atmosphere. With the exception of ultraviolet, and some forms of infrared and radio radiation, a typical atmosphere is thick enough to support a humans and a biosphere, however the exception is ultraviolet light. The most lethal forms can only be blocked by the presence of an ozone layer produced by free oxygen. Without an ozone layer, UV-B radiation reaches the surface and is eventually lethal to exposed, unshielded life.

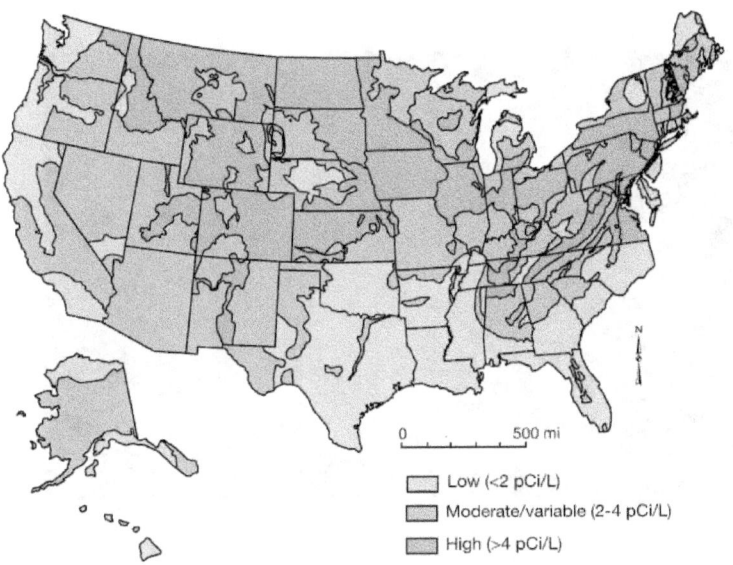

Low (<2 pCi/L)
Moderate/variable (2-4 pCi/L)
High (>4 pCi/L)

Life can still exist in the water, which provides enough shielding for ultraviolet only a few meters below the surface, but land-based life must operate in the night-time and has to be protected during the harsh daytime. Organic optical sensors (eyes or photoreceptors) will quickly form cataracts if unprotected for long periods of time. UV-B is also a known mutagen so prolonged surface exposure leads to the eventual accumulation of harmful mutations unless they can be repaired quickly. Meanwhile, there are no known problems with infrared and radio emissions that reach the ground.

In Space - Here we have to deal with cosmic rays during the voyage, and both cosmic rays and high-energy particles generated in stellar flares once we reach our target planetary system.

Interplanetary Travel

One of the closest sources of cosmic radiation other than our sun is the Crab Nebula; the remains of a massive star that exploded in 1054 AD at a distance of 6,500 light years.

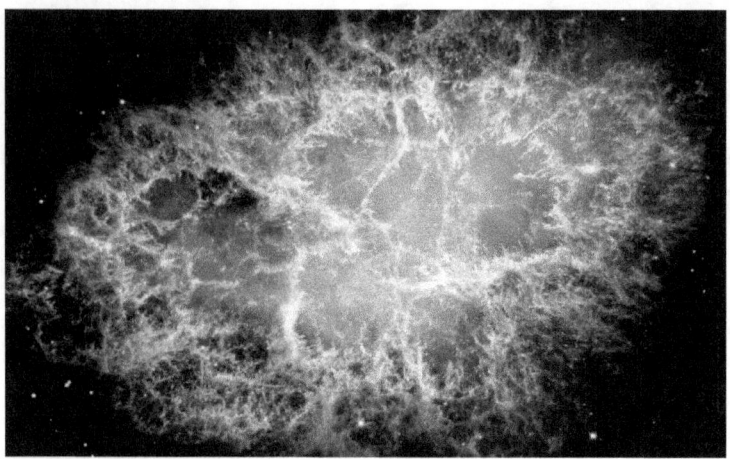

Cosmic rays are far more energetic than other kinds of particle radiations emitted by stars, because they have been generated by the energy of entire exploding stars and then accelerated for millions of years to enormous energies as they circulate around the magnetic field of the entire Milky Way. Cosmic rays from a supernova are like the energy emitted by a camera's flash, but the highest energy cosmic rays are created by a process similar to what scientists use in 'atom smashers' like the Large Hadron Collider.

In deep space far from a star, you will experience the full-bore intensity of cosmic rays. In the solar vicinity cosmic ray intensity is what we call 'isotropic' because it is about the same in all directions. There are no favored directions you can travel that minimize its intensity because cosmic radiation will penetrate your living quarters from all directions. Cosmic rays also come in a variety of different energies. When physicists discuss them, they find it convenient to describe their 'energy spectrum', which gives the flux of cosmic rays at various energies. This instantly tells a physicist how the energy is spread out, and how often you should expect to detect a cosmic ray particles of a particular energy.

This figure, for example, shows that at an energy of '10^{11}' electron volts (eV), you should expect to detect about one particle every second in an area about 1 meter^2. Since 1 eV equals 1.6×10^{-19} joules, the cosmic ray carries about 1.6×10^{-8} joules of energy per hit. This doesn't sound like much, but in atomic terms it is HUGE. It equals 100 GeV, and these kinds of collisions at laboratories such as

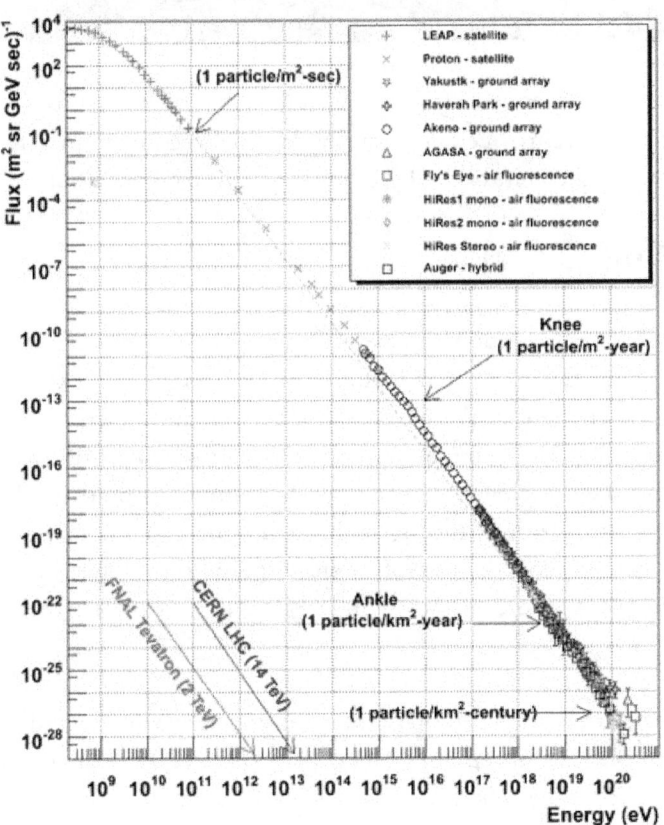

Cosmic Ray Spectra of Various Experiments

CERN and Fermilab lead to rain storms of secondary particles.

So, instead of having to block or shield yourself from just one particle, you have to shield against hundreds of secondary radiation particles after they hit the wall of your ship...or your spacesuit!

Shielding

The intensity of radiation can be reduced in many different ways depending on the type of radiation and its energy. For example, ordinary visible light can be shielded by any opaque substance, but x-rays require a denser medium. Similarly, various forms of matter radiation can be shielded by increasing the number of particles they have to scatter off of as they pass through a medium. Each time a particles of radiation scatters off a shielding particle it loses some of its energy, and the farther it penetrates, the more energy loss occurs.

As a general rule, the more mass you can put between you and the source of the radiation, the better off you are. But there is a problem. For very energetic particles of radiation, when they interact with the shielding material, they can emit secondary radiation particles that, themselves, can be a problem. For the highest-energy cosmic rays, the more shielding you use, the more secondary radiation you generate inside the shielding, so now you have to find an optimum shielding thickness that reduces the dosage from the secondary shielding radiation too! This is why designing the proper shielding for a spacecraft can be a tricky proposition, and you really have to know the energy spectrum of the particles you are trying to shield against to get things right.

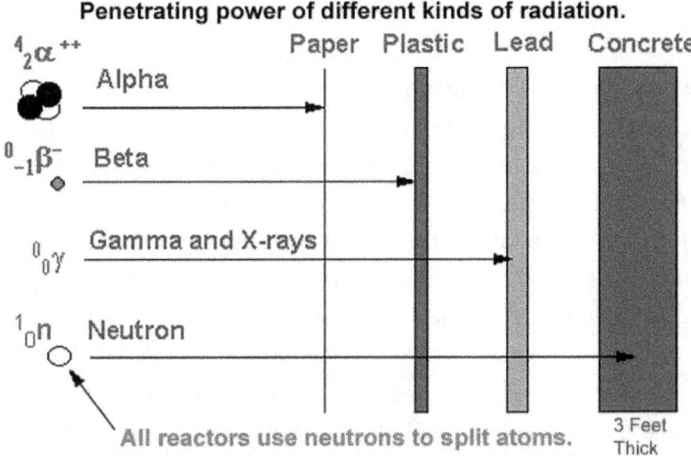

Penetrating power of different kinds of radiation.

Biophysicists who measure and worry about the radiation health effects to humans work with a selection of materials used in designing spacecraft bulkheads, spacesuits and habitation modules to come up with the best design for radiation shielding given the expected radiation levels that the crew is exposed to during a particular period of time. The basic idea is to first decide on what the maximum expected radiation dose can be over the exposure time of the crew, which does not exceed established healthy levels. Once you know the target level on the human-side of the shield, you then predict what the expected environmental-side looks like in terms of types of radiation and their energies. These two numbers tell you by what factor you have to reduce the radiation flux with the shielding material. The last step is to find an appropriate shielding material and thickness that meets the size and mass constraints set by the engineers. Here are some examples:

Spacesuits and EVAs – The amount of shielding is about 0.3 gm/cm² and this is enough to protect an astronaut from most forms of cosmic radiation of moderate energy for up to several hours. However, a major solar flare that occurred on August 2 1972 would have exposed a lunar astronaut to over 500 Rems of radiation in the spacesuit, which would have been life-threatening without prompt medical attention – but not necessarily fatal.

International Space Station – The shielding from the average bulkhead is about 5 grams/cm². The average radiation dose for ISS astronauts working in space for 6 months is 0.05 to 2 Seiverts (5 to 20 Rems). However, on January 20, 2005 a powerful solar flare called NOAA 720 bathed the ISS in high-energy 100 MeV proton radiation for several days. An astronaut on EVA would have gotten radiation poisoned, but the ISS crew inside received only about 1 Rem of exposure thanks to the bulkhead. This is the same exposure they would have received living on Earth in two years, but on the ISS they received the same dose in a matter of a few days.

Trip to Mars – The transit vehicle will probably provide about 10 to 15 gm/cm2 of shielding. The Mars rover Curiosity has allowed us to finally calculate an average dose over the 180-day journey. It is approximately 300 mSv, the equivalent of 24 CAT scans. In just getting to Mars, an explorer would be exposed to more than 15 times an annual radiation limit for a worker in a nuclear power plant. Once on the Martian surface, cosmic radiation coming from the far side of the planet is blocked by the planet's own mass, but you still will get it from the sky. But on the day-side of Mars, protection from strong solar particles varies considerably as the thin Martian atmosphere is bombarded by the solar wind. Although Earth's atmosphere is equal to a shielding of 240 gm/cm2, the thin Martian atmosphere provides only about 12 gm/cm2. That means on the surface, the unshielded radiation dose rate can vary from 10 to 20 Rems/year, which is over 30 to 60 times higher than on Earth.

Cosmic Ray Environment
Dose Equivalent Values (rem/yr)

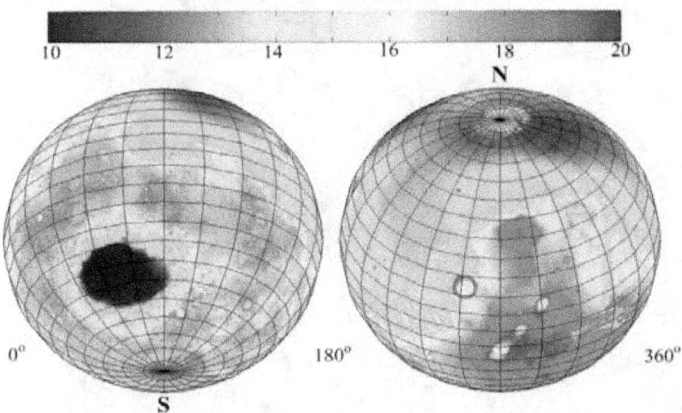

This brings up another issue with radiation: the rate of exposure. One Rem of radiation delivered over a year or more over your entire body surface area and volume can be completely harmless, but the same radiation delivered in a few seconds can be a major hazard. Radiation damages cells, but cells can repair themselves given enough time. The problem occurs when the radiation rate is so high a cell does not have time to complete one repair before the next damaging event happens and the repair system is overloaded. This happens at about five weeks. This sounds like a long time per cell, but at a high enough radiation level, the same cell can be struck by a radiation event multiple times within this five-week time, causing the repair mechanisms to fail resulting in cell death. Cells are not the stable entities you think they are. Every hour over 10,000 measurable DNA changes can happen in a single cell. Thankfully, we have many protective repair mechanisms that

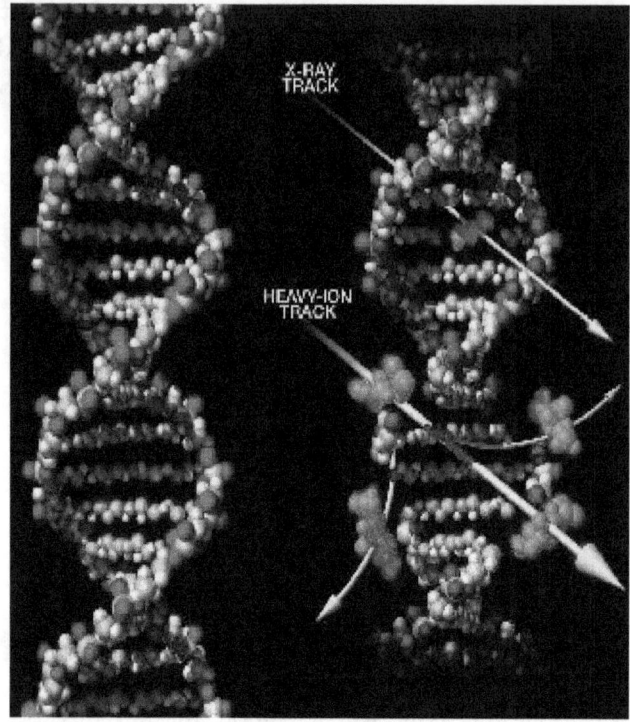

constantly look for damage and then take the necessary molecular steps to return the organism to a functioning state. Under a radiation exposure of 1 centi-Grey (1 Rad) about 100 DNA changes occur. If we convert this in to an annual repair rate, we get about 90 million DNA changes per cell per year from ordinary biological activity, and for the average ground-level dose of 0.24 cGy/yr we get about 24 changes. So under normal conditions, the radiation background is utterly irrelevant.

Overall, the response of an organism to an acute radiation dose or an increase in dose rate is generally described by a dose-response function that changes from low-dose benefit into high-dose harm, at a specific threshold called the NOAEL, which means 'no observed adverse effects level'. For a short-term radiation dose, ~ 50 Rem (0.5 Gy), and for a lifetime dose rate, ~ 0.7 Gy per year.

Solar flares often roil the surface of our sun due to complex magnetic fields, but are harmlessly dissipated and blocked by Earth's atmosphere. This is not the case for astronauts operating in space. This image shows sun erupting with an X1.7-class solar flare on May 12, 2013. It was taken by NASA's Solar Dynamics Observatory.

But perhaps there is something else we can try. Suppose we intercept the particles before they even reach the ship?

Another way that would work for electrically-charged particles is magnetic shielding. This is actually being studied by NASA. Let's have a look at this elegant and very science-fiction-like solution and see where it takes us!

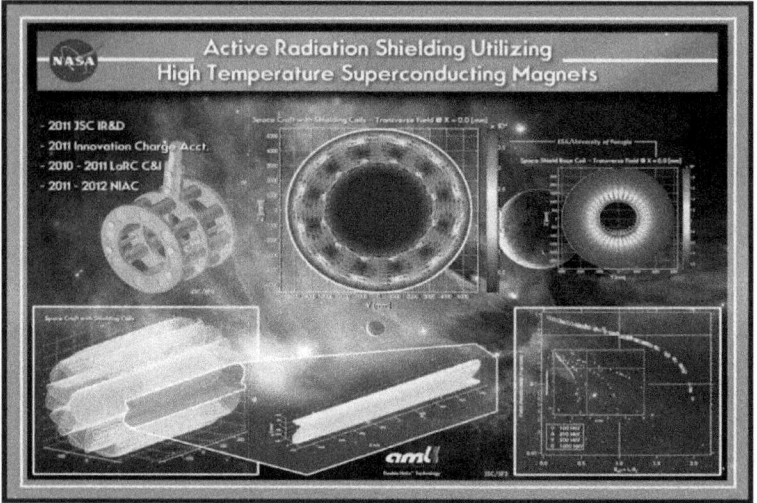

Magnetic fields as shielding

Astronauts can work in the International Space Station because not only is there still some dilute atmosphere to help with the shielding as well as the bulkhead of the space station itself, but because Earth's magnetic field deflects the highest-energy particles away from the inner Low Earth Orbit regions. Nevertheless, this magnetic shielding is not perfect and breaks down for the most energetic cosmic rays, which still make it to the ground. There are also 'air shower' events caused by the highest-energy cosmic rays that produce a steady rain of radiation at the surface. Inside the ISS, astronauts have to rush to the more-protected areas with the thicker bulkheads to avoid the occasional solar flare radiation events.

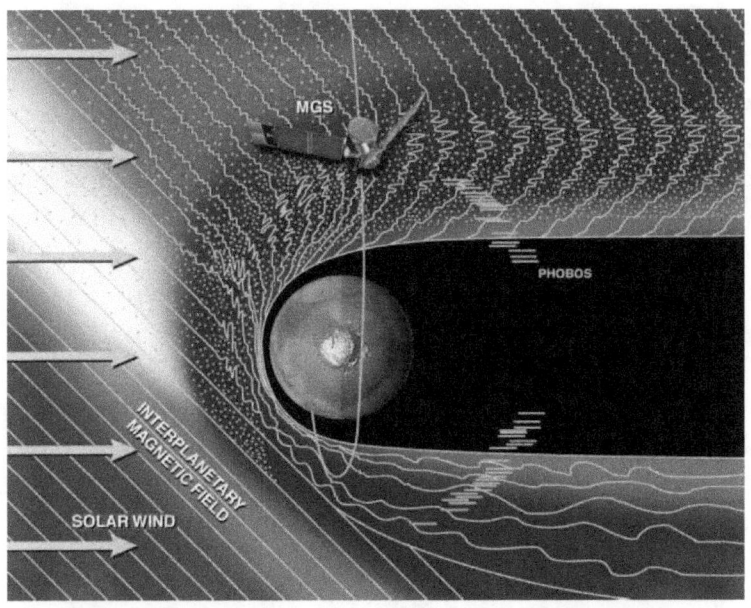

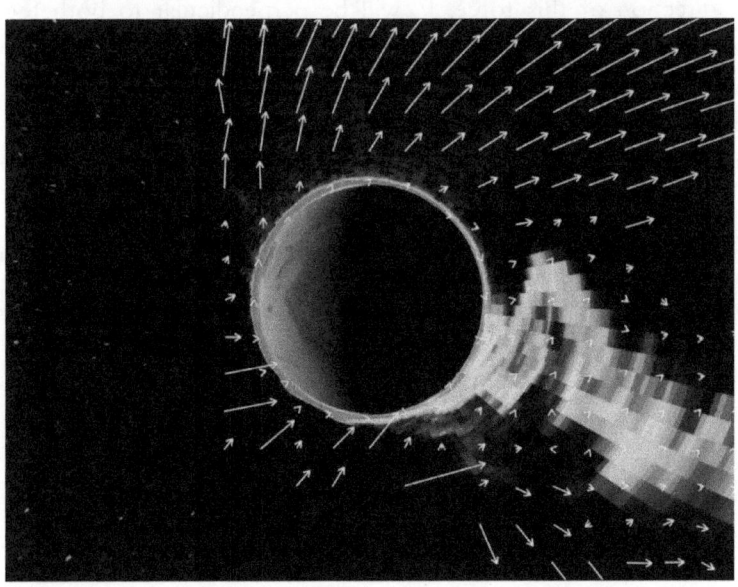

Interplanetary Travel

Here we see a simulation of Mars atmosphere loss due to solar wind. Mars does not have a magnetic field and this probably helped the planet lose most of its atmosphere over billions of years.

How about using magnetic fields to shield the ship from cosmic rays? Magnetic fields have been investigated many times as possible shields, but they always run up against the same problem. The intense field needed to do the shielding is lethal to humans, to say nothing of sensitive mechanical or electrical systems!

To understand magnetic shielding is a bit trickier than the simple ballistics of dust grains. A magnetic field with a strength represented by B, will create a force on a charged particle with charge Q and speed V according to the Lorentz Force Law

$$F = Q \, V \, x \, B$$

The direction of this force, F, will be perpendicular to both the direction of the magnetic field, B, and the velocity, V, of the charged particle, Q. For example, if a charged particle travels along your line-of-sight, and the magnetic field is oriented north-to-south, the force the charged particle feels will be from east-to-west.

Another thing that happens is that the particle feels this as a centripetal force that tries to get it to spiral around the magnetic lines of force. Centripetal forces are calculated by another formula

$$F = MV^2/R$$

where V is the speed of the charged particle and R is what is called the gyroradius. When we put these two formulae together in balance and simplify with a little algebra we get

$$B = MV/QR$$

But the kinetic energy of the particle is $E = 1/2 MV^2$, so in terms of the particle energy we get

$$B = 2E/QVR$$

What this means is that if we want to shield ourselves from a $E = 150$ MeV cosmic ray particle over a thickness of a generous $R = 1$ meter

> If we want a proton with a charge of $Q = 1.6 \times 10^{-19}$ Coulombs to be deflected by $R = 1$ meter traveling at $V = 60,000,000$ meters/sec (20%C) with an energy of $E = 150$ MeV where 1 MeV $= 1.6 \times 10^{-13}$ Joules, we get
>
> $B = 2 (150 \times 1.6 \times 10^{-13})/(1.6 \times 10^{-19} \times 6 \times 10^7 \times 1)$
>
> $B = 5$ Teslas.

spacecraft bulkhead, when you work the numbers in the formula, we need a magnetic field of $B = 5$ Teslas. This doesn't sound like much, but when you consider that this field has to be maintained across the entire surface area of the spacecraft, this presents a severe problem. The total energy required is just $E = 8\pi B^2$ x Volume, so that a 5-Tesla field in a volume the size of a large building (50-meters x 50-meters x 100-meters) requires 1.6×10^{15} joules. This is as much energy as 0.4 megatons of TNT. Fields this strong, though occupying only a few cubic meters, are common in hospital MRI machines. The strongest continuous magnetic fields currently available are about 45 Teslas, and require about 45 megawatts of power to sustain, in a volume less than a cubic meter.

There have been many studies since the 1960's that have attempted to design spacecraft shielding using superconducting magnetic fields. The currents required to shield even a few hundred cubic meters of space are in excess of one million amperes for deflecting 100 MeV protons. The required field strengths are between 5 and 10 Teslas. Because the coils are superconducting, you only have to energize them once for the trip, requiring 55 hours of charging using a 10 kilowatt power source. But the stored energy, if released because the superconducting system failed, would cause catastrophic melting of the spacecraft!

Mutations

Mutations in our genetic code are inevitable because even during the act of replication the billions of base-pairs in our genome can result in occasional misspellings of the A ,G ,T and C coding for specific proteins. These lead to the production of different proteins than required in a particular gene. Usually these errors can be repaired by processes in the cellular nucleus and are rendered harmless. But if this repair process is overwhelmed because you are in a high-radiation environment, mutations can persist. When these happen in ordinary somatic (body) cells, this can either lead to a dead cell, or one behaving badly like a cancer cell. When the mutations appear in a sperm or egg cell, they can be passed on to the next generation. This happens in cases like sickle cell anemia, where a single base-pair error leads to a protein that causes a

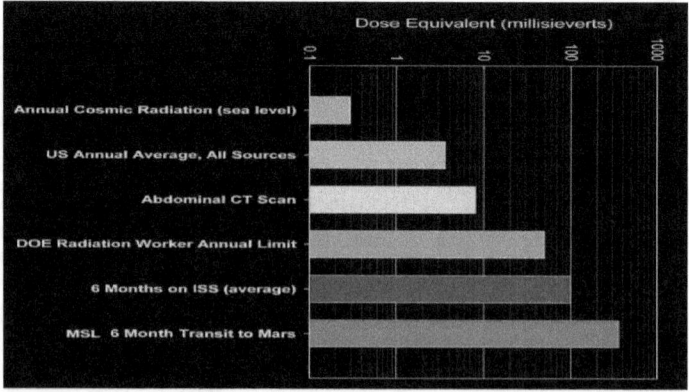

malformed red blood cell.

Because women have two Y chromosomes and men have an X and a Y, genetic errors in the Y chromosome can often be repaired because the corresponding healthy Y chromosome has a good copy of the particular gene. For men, this natural redundancy is not available, and this is why men have more sex-linked diseases that shorten their lifespans than women do. In some sense, women are a better flight crew than men from the standpoint of genetic health.

When NASA's Curiosity rover was en route to Mars on a 253-day voyage through interplanetary space, its radiation sensors directly measured how 'hot' interplanetary space was and whether such a journey would be a hazard for future travelers.

This bar graph compares the radiation dose equivalent for several types of experiences, including a calculation for a trip from Earth to Mars. The data show that during a typical 6 month cruise to Mars the astronaut crews would be exposed to nearly four times the typical 6 month exposure of astronauts aboard the ISS. The spacecraft, though not designed to shield humans, was exposed to an average of 1.8 milliSieverts (180 milliRem) per day during the trip to Mars due mostly to galactic cosmic rays. The round trip dose of 9 Seiverts (90 Rads) doesn't even include the astronaut's surface stay on Mars, which is about 50% less than in space, thanks to what there is of the thin Martian atmosphere. It is hoped that added shielding can bring these dose numbers way down, while at the same time not adding huge costs and weight to the mission.

Radiation does more than 'merely' cause genetic damage. It can also cause an accumulation of brain injuries that over time affect cognition and memory. Researchers used to think that slow-growing brain cells were less vulnerable than fast-reproducing cells to radiation damage, but this now seems to be no longer true.

251

Interplanetary Travel

When high-energy, highly-charged particles enter brain tissue, they cause a narrow channel of cellular damage as they pass through the skull and exit. They also produce secondary particles as they collide with the nuclei of atoms in these brain cells. These secondary particles can take their own paths through the brain and damage other cells over time. According to a recent calculation by Marcelo Vazquez, a researcher at Brookhaven National Laboratory, large fractions of the cells in astronaut's brains would be hit at least once by these energetic particles on a 3-year trip to Mars. Astronauts with severely damaged brain cells could experience certain kinds of memory impairment and also become depressed. They might even develop conditions resembling Alzheimer's disease.

A single base pair mutation in the human genome results in blond hair and blue eyes because of a melanin deficiency, but sometimes the process goes awry. The resulting blue eyes in people with plenty of melanin is especially dramatic, but entirely harmless. Other single base-pair errors are far more problematical such as sickle-cell anemia.

The main problem we have to face with human voyages to the planets and setting up colonies there is that the radiation exposure will inevitably produce mutations in the human genome. How we deal with these mutations medically and socially is a matter for long term concern if we want our future colonies to prosper and remain healthy.

The Hazards of Interplanetary Dust

Contrary to what you might imagine, space is not empty. We know from nightly meteor displays that it is filled by small nuggets of matter. These rain down on the surface of Earth at a rate of 100 to 300 tons of cosmic dust every day. This dust is very small, microscopic in fact, but there is so much of it that it outweighs the occasional golf-ball or watermelon-sized meteor that also pelts our world every day.

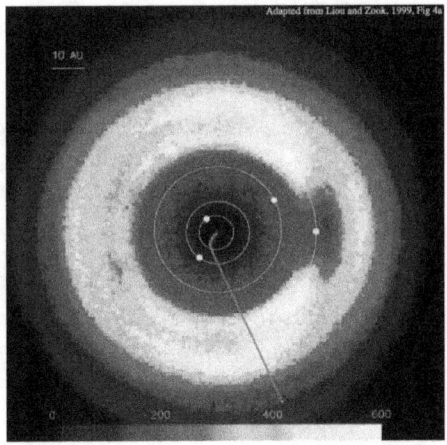

Interplanetary Travel

The Student Dust Counter engineered by students at the University of Colorado, is attached to the New Horizons spacecraft measuring dust on its way to Pluto. This image is a computer simulation of the possible dust distribution in our solar system based on the SDC data. The bar at the bottom shows the number of particles per AU². The green line shows the path of the New Horizons spacecraft. The SDC instrument can detect dust grains with a mass between on billionth and one trillionth of a gram.

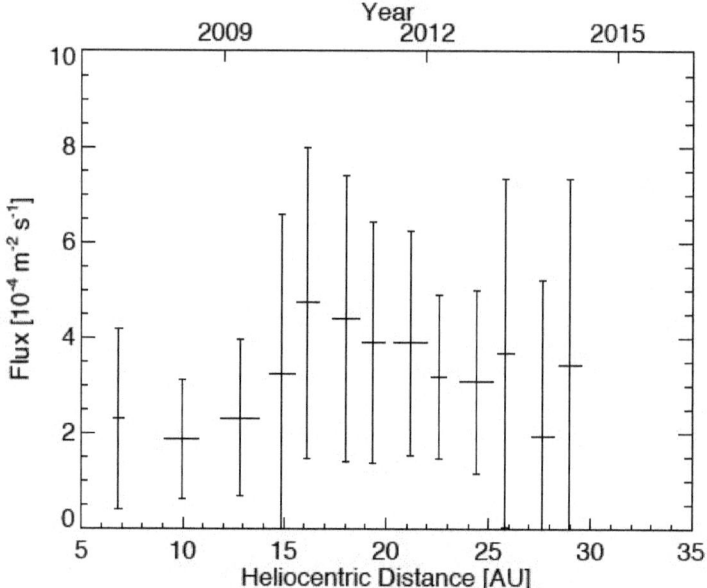

Even so, the figure shows that over the course of the 8-year journey, the New Horizons spacecraft SDC detector, which is about 0.1 meters² was struck by one of these dust grains every week.

These micrometeors are created in the asteroid belt and in the tails of comets, and this dust fills most of the entire volume of the solar system. We can see the glow of this material in reflected sunlight called the Zodiacal Light. Earth plows through this debris and for the larger dust particles the size of rice grains, we get meteors in the night sky. But for the smaller dust grains we see nothing at all because the micrometeors barely make a flash as they penetrate the atmosphere. All of this interplanetary dust would amount to the same material in an asteroid about 30 kilometers across, but the solar system has had billions of years to generate this dust by asteroid collisions, comet evaporation, and some of it even comes from interstellar space!

From 2002 to 2006 the NASA Ulysses spacecraft measured high speed dust grains as they impacted the spacecraft's sensitive dust detector. These dust grains have masses between 10^{-20} to 10^{-11} kg, with most near 10^{-16} kg. Here we see a scanning electron microscope image of an interplanetary dust particle. It's very rough and seems to be aggregates of large numbers of sub-micrometer grains clustered in a random order.

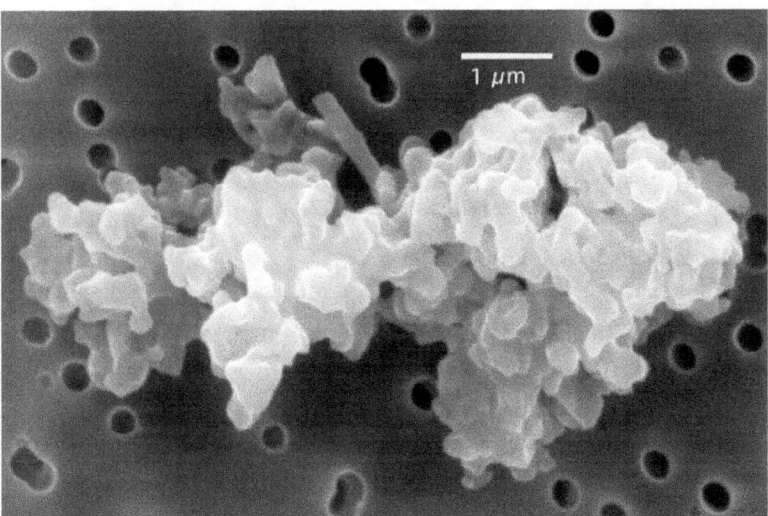

Farther out in space, the Voyager 1 and 2 spacecraft have studied dust impacts as far away as 60 AU from the sun. The spacecraft encountered micron-sized dust particles with masses of about 10 trillionths of a gram, with a density of about 2×10^{-8} particles per cubic meter. The most likely sources for these particles are comets. As the table below shows, both Voyager 1 and Voyager 2 (in parenthesis) noted a sharp fall-off in these dust particles beyond the orbit of Pluto between 30 and 50 AU.

It is hard to really take these miniscule dust grains as a serious hazard for interplanetary travel, but we know that they are only a small part of a much bigger picture of interplanetary debris. The largest forms of this debris are centimeter and meter-sized bodies that exist by the millions. Encountering just one of those in a high-speed interplanetary voyage could be fatal.

The amount of interplanetary 'junk' is not a mystery, but has been studied for decades. On Earth, at the bottom of a gravity well, we get pummeled by particles as small as grains of rice that produce typical nighttime meteors, to things big enough to cause global extinction that are kilometers across. On February 15, 2013 the Chelyabinsk Meteor lit up the skies over Russia as a 20-meter object entered the atmosphere at 19 km/sec. Generally, the smaller the object, the more frequent they are. A 20-meter object may enter the atmosphere every 100 years, but a 1 cm object may enter every few minutes! Here, for example, is a NASA plot of the reported bolide events caused by meteors between 1 and 20 meters in diameter between 1994 and 2013.

On average, 33 metric tons (73,000 lbs) of meteoroids hit Earth every day, the vast majority of which harmlessly ablates ("burns up") high in the atmosphere, never making it to the ground. The moon, however, has little or no atmosphere, so meteoroids have nothing to stop them from striking the surface. The slowest of these rocks travels at 20 km/sec (45,000 mph); the fastest travels at over 72 km/sec (160,000 mph). At such speeds even a small meteoroid has incredible energy -- one with a mass of only 5 kg (10 lbs) can excavate a crater over 9 meters (30 ft) across, hurling 75 metric tons (165,000 lbs) of lunar soil and rock on ballistic trajectories above the lunar surface.

Bolide Events 1994–2013
(Small Asteroids that Disintegrated in Earth's Atmosphere)

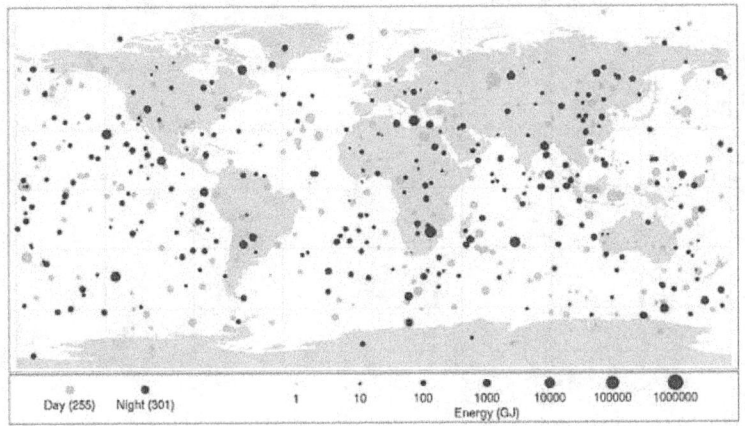

Since 2006, NASA's Meteoroid Environment Office and the Marshall Space Flight Center's Space Environments Team have been monitoring the moon's surface during darkened phases to catch the brief light flashes of meteor impacts. The predicted sizes are between a few grams to several kilograms. Over 20 of these events were spotted in 2013 alone.

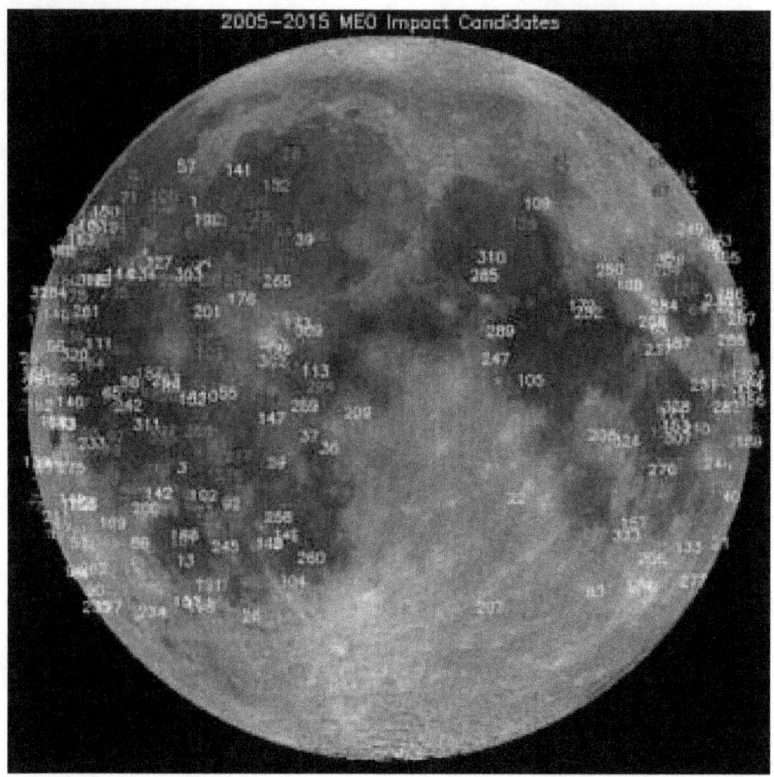

Our Earth is located in what amounts to an interplanetary shooting gallery, with large objects constantly crossing our orbit every day, or coming near enough to be worrisome. Take for example the occasional very bright meteors you see in the sky, called bolides. These can be inches to meters in size, and thanks to the 15 cameras in the NASA bolide network of sky cameras, we can now record many of them every day. By triangulating their paths through the sky, we can work out what orbit they were on as they collided with Earth. Have a look at this recent orbit calculation on August 12, 2015 when 195 fireballs were recorded during the Perseid Meteor Shower during a single day. (Image credit: NASA Meteroid Environment Office).

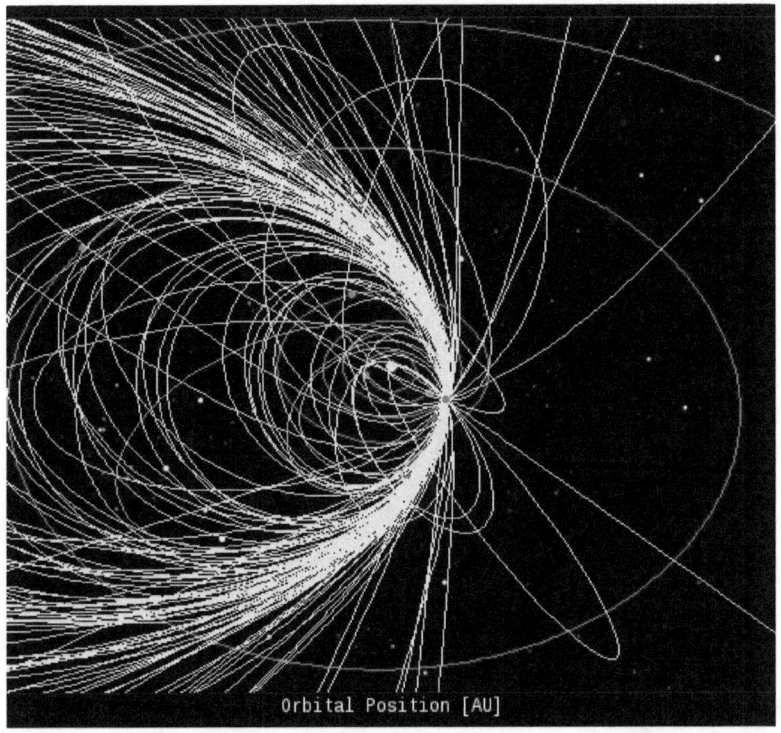

Orbital Position [AU]

The IAU Minor Planet Center also tracks detected small bodies that come close to Earth. This table lists the ones for June 3, 2015 out of 1586 known potentially hazardous asteroids. The distance unit of LD=1 corresponds to the Earth-moon distance of 384,401 km. The vast majority of the close encounters were from asteroids discovered in 2015, which suggests that many more of these close encounters are occurring with small objects under 100-m in size that we have yet to discover. If we had actually found all of them, then there would be about equal numbers of objects discovered each year listed in this table since 1994. (Table from spaceweather.com)

Asteroid	Date(2015)	Miss Distance	Size
2015 KP57	May 28	10.4 LD	44 m
2015 KW120	May 29	1.1 LD	27 m
2015 KH	May 29	14.3 LD	53 m
2015 KQ120	May 31	8.5 LD	20 m
2015 KM57	Jun 3	6.6 LD	36 m
2005 XL80	Jun 4	38.1 LD	1.0 km
2015 KA122	Jun 6	3.3 LD	101 m
2015 KU121	Jun 7	7.5 LD	109 m
2012 XB112	Jun 11	10.1 LD	2 m
2015 KK57	Jun 23	8.3 LD	13 m
2005 VN5	Jul 7	12.6 LD	18 m
2015 HM10	Jul 7	1.1 LD	73 m
1994 AW1	Jul 15	25.3 LD	1.4 km
2011 UW158	Jul 19	6.4 LD	565 m
2013 BQ18	Jul 20	7.9 LD	38 m
1999 JD6	Jul 25	18.8 LD	1.6 km
2005 NZ6	Aug 6	76.5 LD	1.4 km

In interplanetary space, numerous spacecraft have now plied the vast volumes stretching from Earth orbit all the way out to Pluto. No spacecraft have been lost from a wayward impact, but instruments have recorded a constant pelting over the months and years of these typical flight times. From this data, scientists can put together a frequency graph that shows about how often objects of various sizes will impact a square-meter of surface over time.

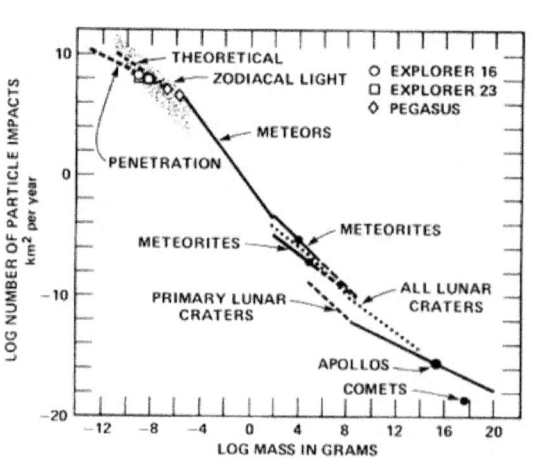

For a spacecraft with a 500 meter2 surface area in space, you might expect to be struck by a 1 gram meteor every 2000 years! The problem is, what level of risk do you feel is appropriate if one collision can cause severe, perhaps even fatal damage? Can you see them coming in enough time to divert your course to avoid them?

> From the graph for Log(M) = 0 the vertical axis says log N = -2 so 1 gram objects strike at a rate of 10^{-2} per km^2 per year. The spacecraft surface area is 0.05 km^2, so the expected rate is 10^{-2} x 0.05 = 0.0005 per year. Your spacecraft has a 1 in 2000 chance of being hit.

Currently there are many sophisticated radar systems that can detect millimeter-sizes objects in Earth orbit. NASA's main source of data about orbital debris in the size range of 1 to 30 cm is the Haystack radar. The Haystack radar, operated by MIT Lincoln Laboratory, has been collecting orbital debris data for NASA since 1990 under an agreement with the U.S. Department of Defense. Haystack statistically samples the debris population by "staring" at selected points in the sky and detecting debris that fly through its field-of-view. The data are used to characterize the debris population by size, altitude, and orbit. From these measurements, scientists have concluded there are over 500,000 debris fragments in orbit with sizes larger than one centimeter.

The 70-meter Goldstone antenna located near Barstow, California, when operated as a 400-kilowatt radar, is capable of detecting 2 millimeter debris at altitudes below 1,000 km. It has also been used to detect and

study the moons of Jupiter and Saturn. Its maximum range is the orbit of Saturn for objects hundreds of kilometers across.

All of these installations are massive dishes, or equally large phased-arrays many tens of meters across, that generate megawatt pulses of radio-wave energy. If you are only interested in meter-sized objects, ESA has recently built and tested a small radar system for collision avoidance on the International Space Station and on other satellites. It can detect meter-sized objects at a distance of up to 500 km. Meanwhile, the Arecibo radio telescope 300-meters in diameter located in Puerto Rico, is routinely used to study and image asteroids in the Asteroid Belt that are tens of kilometers across, and it can detect objects tens of meters across out to the orbit of the moon.

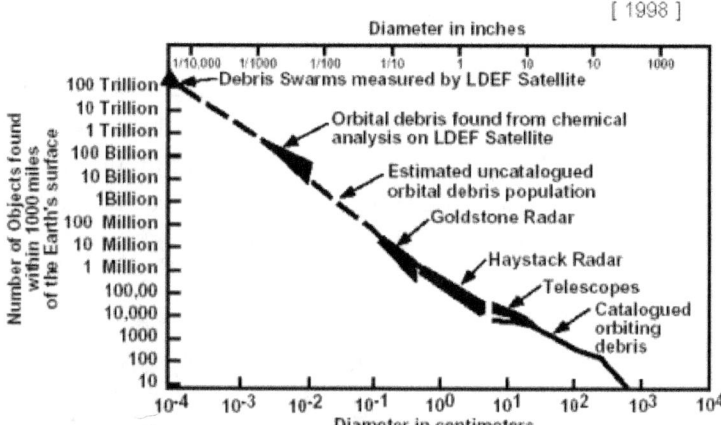

As these examples show, radar detection is a complicated balance between object size, detection range, and the transmitted power of the radar pulse. At a distance of a thousand kilometers, for a ship traveling at a healthy speed of 50 km/sec you only have about 20 seconds before collision. In order to have an hour's notice, you have to detect the object at a distance of 180,000 kilometers or nearly the distance to our moon from Earth.

Here is an example of a half-inch hole that was made in the Space Shuttle Endeavour's radiator panel by the impact of unknown space debris during STS-118 in 2007. The impact speed was typical of orbital debris of 10 km/sec. Space Shuttles and the ISS are constantly bombarded in Earth orbit by paint chips and other small debris only a few millimeters across, but traveling at 28,000 mph they are worse than common bullets.

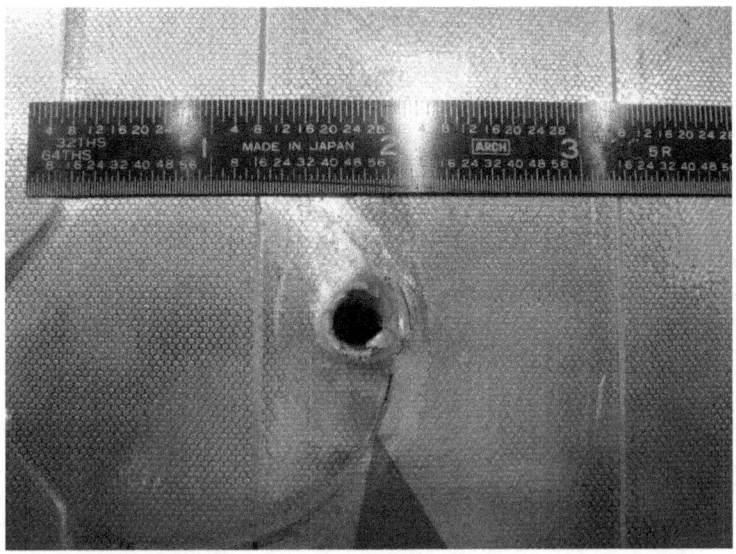

Orbital debris is a hazard for catastrophic mission failure, and is one of the major worries of the International Space Station, whose large solar panels collect hundreds of collisions every year, degrading the performance of this vital electrical system over time. Although the near-Earth debris cloud is not directly relevant to interplanetary voyages, it does point out how even rare but small impacts can have devastating effects on the health and safety of such long-duration missions traveling through space, populated by unknown and undetectable hazards.

There is no technology on the drawing boards that helps interplanetary spacecraft avoid the lethal effects of dust/asteroids impacts at very high speeds. The faster you go, the more lethal are the stray dust grains and other cosmic junk, and the less time you have to react to the threat. There is no Earth-based technology that will let us detect meter-sized lethal bodies in space at distances of billions of kilometers allowing comfortable hours of warning time. Spacecraft will have to be constantly vigilant for these impacts, especially along the direction of travel where approach speed is highest, and unfortunately where the impact time is the shortest!

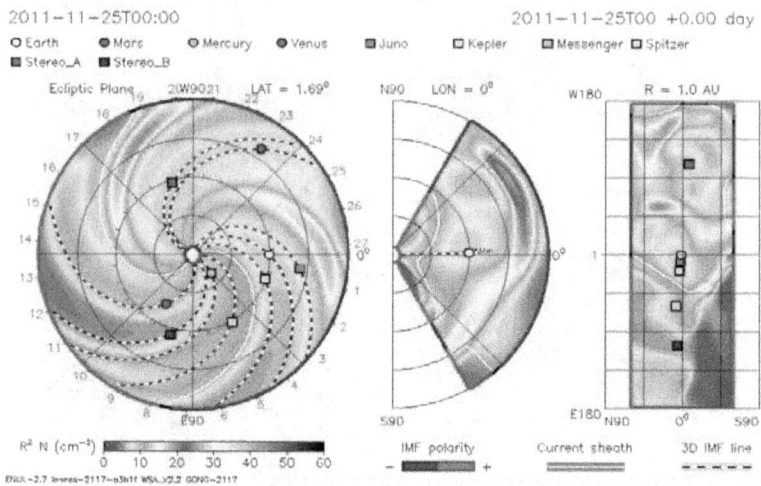

In addition to dust grains, there is also a thin, dilute plasma called the solar wind that travels at up to 400 km/sec leaving the sun. This is mostly made of individual protons and scattered atomic nuclei ejected from the solar surface in a constant stream, and rarely amounts to more than a few dozen particles per cubic centimeter. In its spiral path into the depths of the solar system, it brushes by the planets and compresses their magnetic fields. This is noticeable for Earth, Mars and Mercury, but Venus has no magnetic field. Instead, the wind makes direct contact with the atmosphere creating a comet-like shape as the Venusian atmosphere is dragged behind the wind. It takes almost a full year for the solar wind to reach the outer limits of the solar system where it finally slams up against a billion kilometer shock front as it meets the interstellar medium itself. Fortunately, the dozens of atoms of this wind per cubic centimeter have no important affects in spacecraft traveling through this material. That's one potential hazard we can eliminate!

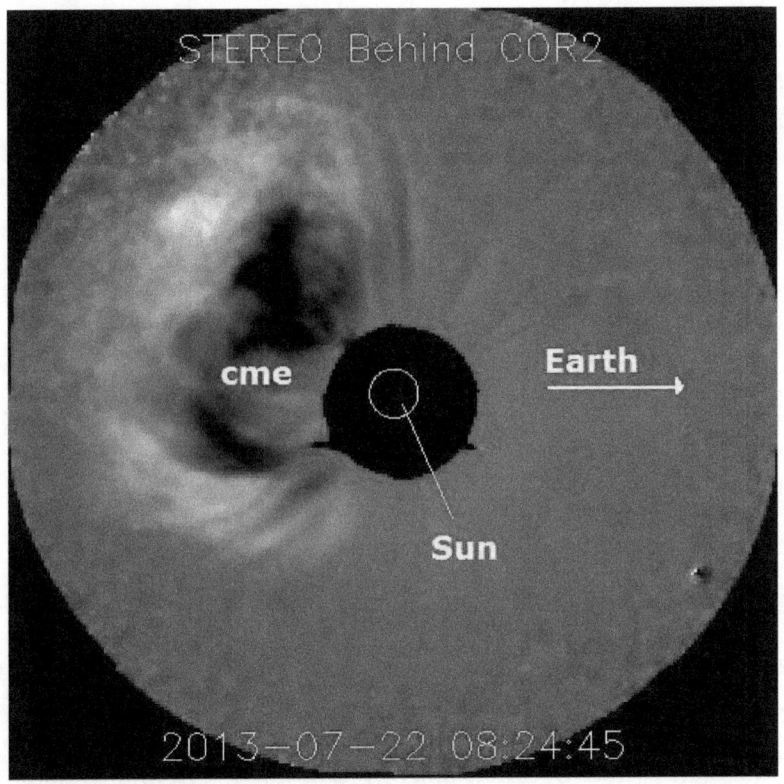

A coronal mass ejection (cme) enters the solar wind (Credit:NASA/STEREO)

Lessons Learned from Antarctica

Antarctica is remote, cold, and remains a continent full of promise, but largely uninhabited some 100 years after it was first traversed.

Antarctica is the coldest of Earth's continents. The coldest natural temperature ever recorded on Earth was −89.2 °C (−128.6 °F) at the Soviet (now Russian) Vostok Station in Antarctica on 21 July 1983

Several governments maintain permanent manned research stations on the continent. The number of people conducting and supporting scientific research and other work on the continent and its nearby islands varies from about 1,000 in winter to about 5,000 in the summer, giving it a population density between 70 and 350 inhabitants per million square kilometres (180 and 900 per million square miles) at these times. Many of the stations are staffed year-round, the winter-over personnel typically arriving from their home countries for a one-year assignment.

There is no economic activity in Antarctica at present, except for fishing off the coast and small-scale tourism, both based outside Antarctica.

Although coal, hydrocarbons, iron ore, platinum, copper, chromium, nickel, gold and other minerals have been found, they have not been in large enough quantities to exploit. The 1991 Protocol on Environmental Protection to the Antarctic Treaty also restricts a struggle for resources. In 1998, a compromise agreement was reached to place an indefinite ban on mining, to be reviewed in 2048, further limiting economic development and exploitation. The primary economic activity is the capture and offshore trading of fish. Antarctic fisheries in 2000–01 reported landing 112,934 tons.

Small-scale "expedition tourism" has existed since 1957 and is currently subject to Antarctic Treaty and Environmental Protocol provisions, but in effect self-regulated by the International Association of Antarctica Tour Operators (IAATO). Not all vessels associated with Antarctic tourism are members of IAATO, but IAATO members account for 95% of the tourist activity. Travel is largely by small or medium ship, focusing on specific scenic locations with accessible concentrations of iconic wildlife. A total of 37,506 tourists visited during the 2006–07 Austral summer with nearly all of them coming from commercial ships. The number was predicted to increase to over 80,000 by 2010.

Cruises to Antarctica:
- Basic Cruise (9-12 days) - $3600 - $9000
- Mountains, Icebergs & Glaciers (11-16 days) - $9000 - 16,000
- Falklands & South Georgia (15-21 days) - $10,000 - 23,000
- Great Explorers Trip (25-29 days) - $12,000 - 35,000

Flights to Antarctica:
- Fly/Cruise Combinations (7-10 days) - $6,995 - 14,995
- Fly Options (1 day) - $18,000 - 23,000

Expedition Options:
- Fly/Camp at South Pole (7 days) - $45,000
- Ski Last Degree (13 days) - $60,000

Basic Rocketry

Thrust

A rocket develops thrust by throwing mass out the engine as fast as possible at a steady speed. The thrust force it develops is the product of the mass rate times the ejection velocity. Rocket engineers in the US measure thrust in pounds of force, but physicists use Newtons. The mass rate Mdot is measured in kilograms per second and the speed V is in meters per second, so

Thrust (Newtons) = Mdot (kg/sec) x V(m/s)

Aerojet engine test (Image credit: NASA)

For example, the J-2 engine on the Saturn V rocket delivered 1 million Newtons of thrust. The fuel was ejected at a speed of 4100 meters/sec, so that means Mdot = 1 million/4100 = 243 kilograms/sec was the engine exhaust mass rate!

The Rocket Equation

Now as the rocket loses the mass it throws out the engine, its total mass decreases, but since the ejection speed stays the same. That means its speed has to increase because of the Conservation of Momentum. At some point you completely run out of fuel, hopefully by the time you reach orbit! A simple equation lets you calculate how massive the rocket has to be to reach a given speed.

$$V = V(exhaust) \times Ln(M/m)$$

The total mass of the payload plus rocket and fuel is M, the mass of the payload is m and the rocket engine ejection speed is V(exhaust). For example, to reach orbit, suppose the final speed has to be 11 km/sec, and the exhaust speed is 4 km/sec. If the mass of the entire (Saturn V) spacecraft is 2,970 tons, the payload mass, m, delivered into orbit by this one-stage rocket is just 190 tons. That means for this engine, the percentage of payload mass was just 6.4%. The rest of the rocket mass was fuel and support structure for the fuel tanks and fuel pumps. In fact, no matter what the total rocket mass, you still get only 6.4% of this mass as payload. The problem is in the engine with its low exhaust speed. Unfortunately, these low exhaust speeds are typical of most chemical rockets, so we are stuck with only sending into Earth orbit a small fraction of a rocket's mass as payload mass. The good news is that the moon's escape speed is only 2.4 km/sec, so for the same rocket engine, the payload percentage grows to whopping 55%, and for Mars with an escape speed of 5.0 km/sec, the same rocket allows for a payload mass that is 28% of the rocket mass.

Specific Impulse

Specific impulse (Isp or SI) is a measure of how effective a rocket engine is at producing an exhaust speed of V against a gravitational acceleration of g, and is defined as

$$Isp = V/g$$

For example, the J-2 engine used on the Saturn V rocket used liquid hydrogen and oxygen, and had an exhaust speed of 4000 meters/sec at the surface of Earth where g = 9.8 meters/sec^2. The Isp of the engine is then Isp = 4000/9.8 = 408 seconds. By comparison, a model rocket engine such as the Estes E9-6 has a specific impulse of 86 seconds. Here are some other rocket technologies and their Isps.

Type	Isp	Duration	Exhaust speed (km/s)	Maximum delta-V (km/s)
Solid	255	minutes	3	7
Liquid-fuel	450	minutes	4	9
Nuclear Thermal	1,000	minutes	9	>20
Pulsed Plasma	2,000	years	20	
Mass Driver	3,100	months	30	
Hall Effect	5,100	years	51	>100
Project Orion	10,000	days	100	60
Ion Thruster	21,000	years	210	>100
VASIMR	31,000	months	300	>100
Solar Sail	76,000	months	750	40

Because the exhaust velocity is just V = Isp/g, the rocket equation can be written as

$$V = Isp \times g \times Ln(M/m)$$

270

So in terms of the final velocity v, the higher the Isp, the smaller the mass ratio M/m has to be, and the less propellant you need. So percentage-wise, the more payload you can place into orbit. Another way to look at this is that the higher the Isp, the faster a rocket will travel, and the shorter will be its travel time for a given payload percentage.

So far, we have discussed a few basic ideas about rockets, but what about destinations? What does it take to actually get into space from a planet's surface and travel from place to place? Usually the first part of this discussion, planet lift-off, is described in terms of escape velocity. A rocket launched from Earth's surface is said to have to reach a certain critical speed before it can 'break free' of Earth's gravity and go into orbit. This is of course a poor, but popular, choice of words. A more exact way to describe it is in terms of an energy ladder. To climb up the gravity energy ladder into space you have to go up a particular number of rungs that is different for each of the orbits you want to achieve.

A Simple Mathematical Model

We can combine all the mathematical tools from the previous chapter and put together a simple model that lets us see how travel times to various objects changes as the rocket Isp and payload-to-fuel ratio changes for various technologies. The beauty of the calculation is that we don't even need to know the specifics of the rocket technology, although in the next chapters we will explore some of the ones currently in use and being developed.

We are going to consider a set of scenarios based upon a 100-ton payload reaching the planet Mars located at a reference distance of 74 million km from Earth. Why 100-tons? That's similar to the mass of the International Space Station in which up to six astronauts appear to be comfortable and have the water, food and recreation resources they need.

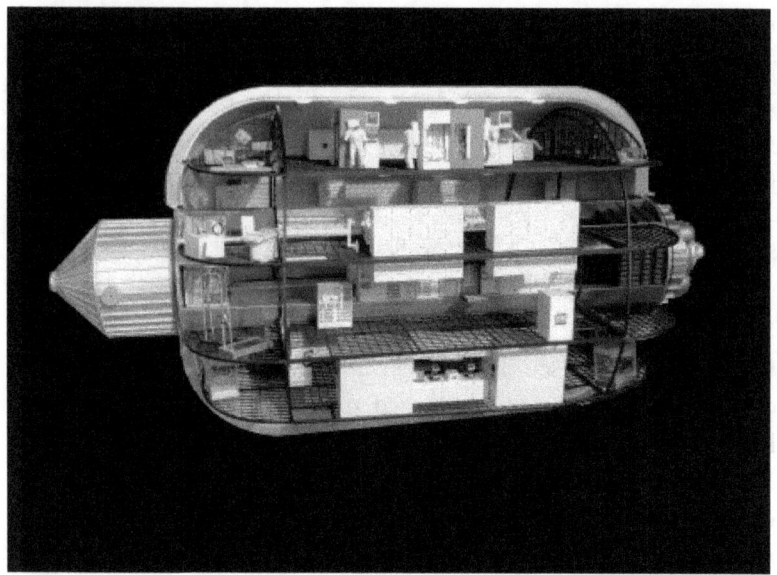

The BA2100, or Olympus, is a conceptual design for an expandable space station module, or interplanetary human transport module, developed by Bigelow Aerospace. It has a volume of 2,250 cubic meters, nearly three times as large as the entire habitation volume of the International Space Station (837 m³). The mass of the BA 2100 could be as low as 65 to 70 tons, but would more likely be in the range of 100 metric tons.

Scenario 1. Let's begin with a rocket engine that delivers an Isp of 1000 seconds. This means the exhaust velocity is 9.8 km/sec. This is about three times the maximum for chemical rockets, but lower than other 'exotic' technologies such as ion or nuclear propulsion. Let's also suppose that we combine enough of these rocket motors together to produce a sustained thrust of 100,000 Newtons. That means the spacecraft has to eject mass at a rate of about (100,000 N)/(9800 m/sec) = 10.2 kg/sec during the entire time the engines are operating. We are going to assume that our spacecraft uses exactly half its fuel to accelerate to its maximum speed and half its fuel to decelerate to reach its destination. The total trip time will be just the combination of how long it takes to use up its one-way full fuel reserves, and it will be rounded to the nearest day. That means that every day the spaceship is losing about (10.2 kg/sec x 86400 sec/day =) 880 tons of fuel. To reach Mars, the ship will need 250,000 tons of fuel to start with. (100 times the Saturn V fuel supply) It will use half of this to reach its maximum speed of 6.8 km/sec after 142 days, and then it will have to turn around and run its engines continuously for another 142 days to get back to zero speed. The acceleration would only be about 0.0008 m/sec^2 so the travelers will be 'weightless' the entire time. The total travel time is 284 days to Mars. This is not a very compelling scenario because of the low-thrust and huge fuel mass involved.

Example of a spacecraft dragging its massive fuel supply! (Image credit: Kerbal Space Program)

Scenario 2 – Let's increase the Isp to 5000 seconds and see what happens! This is not attainable by chemical rockets, but it is a typical level of performance for advanced ion or nuclear technologies. Here we immediately see the benefit of increased engine performance. To reach Mars, the engine would run continuously for 27 days to reach a top speed of 31 km/sec, ejecting 175 tons of mass every day. It would have consumed half of its 10,000 tons of fuel at this point. It would then turn around and decelerate for another 27 days to consume the rest of its fuel, for a total one-way trip time of 54 days. This is starting to look like a decent spacecraft with which to get to Mars and still not risk too many health effects from radiation and microgravity along the away. But Jupiter is still out of reach for a payload mass of 100-tons. To reach Jupiter, this spacecraft would need over 86,000 tons of fuel to maintain a 100,000-Newton continuous thrust, and it would take 243 days for a one-way trip with the usual turn-around maneuver at the 310 million kilometer mid-way point.

Scenario 3 - Well, if you can build one rocket engine that delivers 100,000 Newtons of thrust at an Isp of 5000 seconds, why not use a cluster of these engines to increase the total thrust by, say, ten times to 1 million Newtons? This is what we did in the 1960s for the 2,900-ton Saturn-V rocket where five F-1 engines were combined in a cluster to deliver the required 34 million Newtons of total thrust. When you model this combination, you are ejecting mass at a rate of about 1800 tons/day. Amazingly you need 100,000 tons of fuel and reach Mars in 56 days, which is about the same as for Scenario 2! At the 37 million kilometer midway point on Day 28, your still-very-massive 50,100-ton spacecraft is traveling at 33 km/sec at an acceleration of 0.02 m/sec^2. Let's try a lower fuel mass!

With this higher-thrust engine and 10,000 tons of fuel (about three times the mass of the Saturn V) you burn through half of it in 3 days but only reach a distance of about 4 million km at a top speed of about 35 km/sec. Now, you have the option of coasting at this speed for 22 days then burning the remainder of your fuel for another 3 days to get to Mars. The total one-way trip would be 28 days. So, opting for the 1 million-Newton engine and 10,000 tons of fuel can get you to Mars in 28 days <u>with coasting</u>, while the much heavier ship with 100,000 tons of fuel and constant thrust gets you there in twice the amount of time because there is more fuel mass to push! This demonstrates that for interplanetary travel involving the transport of large payload masses, you need to design more efficient engines with higher Isps to reduce the fuel mass. Just putting clusters of inefficient engines together to get more thrust does not 'beat the curve' and get you there faster because you have to drag along more fuel mass to push.

Scenario 4 – If we explore the consequences of including a coasting phase in the travel, we can get by with smaller fuel masses. Let's look at a design with an Isp of 30,000 seconds, ejecting 14 tons of fuel per day to deliver a modest thrust of 50,000 Newtons with only 3,000 tons of fuel. It uses half its fuel to reach a speed of 194 km/sec after 102 days, at which point it is 768 million km from Earth. We have already blown past Mars (74 million km) and Jupiter (622 million km) during the acceleration phase, and Saturn will be reached before the deceleration phase has completed. This makes Uranus our first port-of-call (2.7 billion km) which requires a 69-day coast period for a total trip time of 373 days. Pluto (5.7 billion km) requires a coast phase of 250 days plus another 102 days to decelerate for a total one-way time of 454 days.

Scenario 5 – Let's crank up the Isp to 50,000 seconds and operate the engine at 100,000-Newtons and see what kind of spacecraft performance we get! The engine's 17 tons/day exhaust is now moving at an impressive 490 km/sec. By Day 4 you are half-way to Mars and have used half of your 140-tons of fuel, reaching a speed of 160

km/sec. Your total ships mass is only 240 tons at the start (42% payload!), and the total one-way time has now shortened to about 8 days! Now that's more like it! To get to Jupiter in 56 days, you need 1000 tons of fuel (payload = 9% similar to the Saturn V!) and your top speed will be 288 km/sec at turn-around. Your acceleration is only 0.1 m/sec^2 so you are weightless throughout the journey just as you were in getting to Mars. Dare we try for Saturn? That would be a 106-day, one-way trip with about 1,900 tons of fuel. Your top speed would be about 310 km/sec. Uranus? That would take 210 days and require about 3,700 tons of fuel! By comparison, the Saturn V rocket required about 2700 tons of fuel and operated for 8 minutes! Now we are talking about spaceships that look as compact and sleek as the ones in science fiction art.

Scenario 6 - None of these designs, as impressive as their travel times are, required accelerations higher than 0.1 Earth gravities, so astronauts remained weightless for the rides. What would it take to reach accelerations of 1 G so our passengers are comfortable the whole way? It requires a combination of more thrust and less fuel mass. One combination might be Isp=90,000 seconds, thrust of 30 million Newtons (61 tons/hour) and fuel mass 6,000 tons. During the 25-day acceleration phase the acceleration varies from 0.5 to 1.0 Gs which is comfortable. You reach a speed of 500 km/sec at a distance of 25 million km, and following a 30-minute coasting phase under weightless conditions, you again have to decelerate. But unless you throttle back on the engine, you will experience from 1 to 4-Gs which is NOT comfortable. This is because the ship now has about half its original mass, but if you keep the thrust the same, the accelerations will be twice as high. So, by running the engines at a thrust of 15 million Newtons, you only experience about 0.5 to 1.0 Gs, and after another 25 days arrive at Mars. The total one-way trip would be about 50 days long, with a very fun 30-minute weightless period in the middle during the turn-around point. This would actually be a comfortable tourist trip, but you still better have things to do to occupy your nearly 2-month journey!

The bottom line is that we cannot create 'artificial gravity' by relying on the ships constant thrust because the required fuel masses even for very efficient engines would be too high. Interplanetary travel will be a weightless activity for all potential rocket technologies with Isps below 100,000. That means we either rotate the ship to create artificial 'centripetal acceleration', or we just sit back and enjoy the ride and hope our bodies won't be in too bad a shape when we get to our destinations!

The best strategies that do not tax our engine technologies involve a short acceleration phase and deceleration phase together with a prolonged coasting period in-between and hopefully at speeds well above 50 km/sec. This allows us to use far less fuel than if we maintained constant thrust for very large payloads. This strategy is reversed for small payloads where small, but constant thrusts are acceptable since the total payload masses are smaller, such as the Dawn, NEAR and Deep Space-1 spacecraft.

We have described some interesting mathematical scenarios that involve engines of differing Isp, but we have not really worried much about the rocket technology that allows for the required Isps. Now it's time to pay the piper and talk detailed rocket engineering. The basic challenge is, how can we design an engine that gives us the highest possible mass flow and exhaust speed (thrust), which translates into the highest Isp? What do we need to do to get Isps above 10,000 seconds for some really superb travel possibilities?

Chemical Rocket Engines

The basic idea in a rocket engine is that the rocket throws mass out its back end as fast and as much as possible. This momentum flow in one direction per unit time is a force. Newton's Third Law says that for every action there is an equal and opposite one, so the rocket exhaust momentum produces a force that drives the rest of the rocket forward in the opposite direction to the exhaust flow. To get the highest thrust you need engines that have the largest Isps.

How does a rocket engine work? There are two typical kinds: Solid fuel and liquid fuel. Within these two main groups there are many variants depending on the chemistry of the fuel.

Solid Fuels

The Space Shuttle used two solid rocket boosters (SRBs) that were shed after they helped the Shuttle reach a desired velocity. They each produced an amazing 13,800,000 Newtons of force. They were 150 feet tall, 12 feet in diameter, and were packed with ammonium perchlorate, which produces an Isp of about 285 seconds. The SRB interior is not a solid plug of fuel but is channeled vertically to maximize its burning surface so that it burns and provides thrust in exactly the right way over its 127-second burn time. It produces about 31 megaJoules/kg in energy.

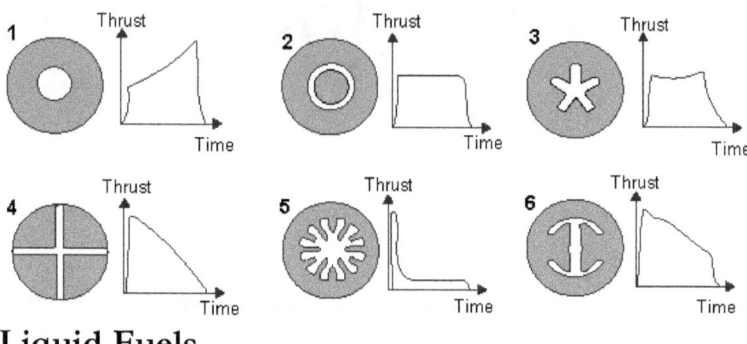

Liquid Fuels

These require a binary system of a fuel and an oxidizer. The liquids are separately pumped at high speed into a combustion chamber whose end has a bell-shaped nozzle. The explosive expansion of the reactants produces high-speed flow of material out the rocket nozzle, which produces the momentum change. This part is similar to the SRB except that liquid fuels can be adjusted in real-time much as you would put your foot on the accelerator of your car. SRBs have only one state, that of being Full On, with thrust controlled by the geometry of the combustion surface inside the SRB core axis.

Because of the huge momenta that these chemical rockets can produce in a short time, they are the only current mean to launch from a high-g planetary surface and reach orbit.

The J-2 engines on the Saturn V used a turbo pump operating at 27,000 RPM to move the liquid hydrogen (LH2) fuel. The liquid oxygen (LOX) oxidizer turbopump operated at 8,700 RPM. The

Isp was 421 seconds. This produces a thrust per engine of 1.3 million Newtons. Mass is discharged at speed of over Isp/g = 4100 meters/sec, and at a rate of 1.3 million Nt/4100 = 317 kg/sec.

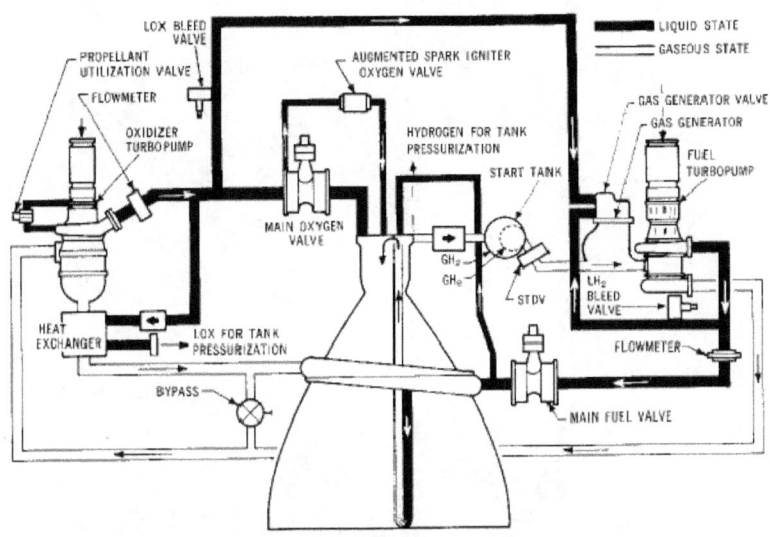

Diagram of J-2 engine used on Saturn V. (NASA)

The highest Isp for a chemical rocket was a wicked, tri-propellant rocket that used lithium, fluorine, and hydrogen as a reaction mass. In 1968, Rocketdyne Corporation was able to achieve a specific impulse of over 540 seconds, but it was a very hazardous mixture. The logistics of putting that in a vehicle are horrendous. LH2 is near absolute zero, and lithium needs to be at a high temp to be a flowing liquid. Not to mention the exhaust products are hideously toxic, and fluorine reacts with almost anything.

Shuttle launch with its SRBs lit up! (Image credit: NASA)

Slow-speed travel

To put all of the proposed ideas on a common footing, we will not worry about how the spacecraft travels that first or last 100 km from the ground to orbit, or from orbit to ground. The part of the journey that takes up the most time is always the travel segment between the parking orbit around Earth, such as LEO or GEO, and the parking orbit around the destination.

We have a selection of likely destinations across the solar system, and some ideas what to do when we get there. How long will these various trips take?

The slowest trips are promised by using chemical rockets with Isps below 500, that also take advantage of gravity assists. This is a popular system used by interplanetary spacecraft such as Galileo, Cassini and the Voyagers. Because the delta-V capabilities are very limited, spacecraft use what are called Hohmann Transfer Orbits to get from place to place by expending the minimum amount of energy and delta-Vs.

Another possibility is to take an even slower route that uses far less fuel. Here we see an artist's concept (Image credit NASA/JPL) of the Interplanetary Transport Network. The thin ribbon represents one possible path from among the infinite number possible within the larger bounding tube connecting each planet and moon. Constricted areas represent locations of Lagrange points. This stylized depiction of the ITN is designed to show its (often convoluted) path through the solar system. Locations where the ribbon changes direction abruptly represent trajectory changes at Lagrange points, while constricted areas represent locations where objects linger in temporary orbit around a point before continuing on.

Interplanetary Travel

The Lagrange points have the peculiar property of allowing objects to orbit around them, despite lacking an object to orbit. While they use little energy, the transport can take a very long time.

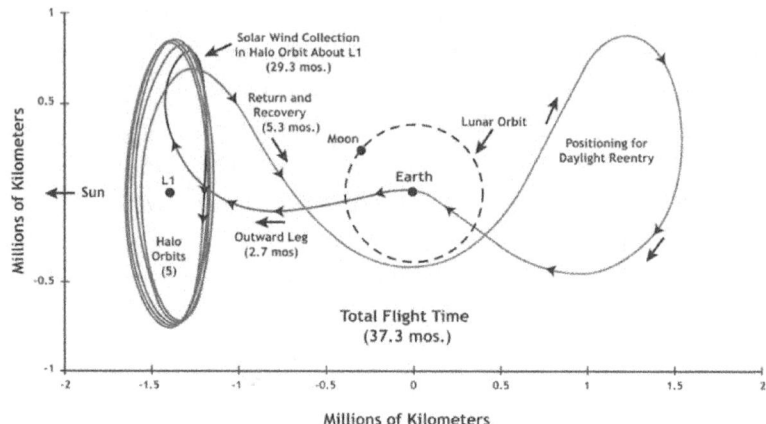

A "freeway" through the solar system can slash the amount of fuel needed for future space missions. Called the Interplanetary Superhighway, the system was conceived by Martin Lo, whose software was used to help design the flight path for NASA's Genesis mission, which is currently using this "freeway in space" on its mission to collect solar wind particles for return to Earth. (Image credit:Wikipedia/NASA/JPL)

Each planet and moon has five locations in space called Lagrange points, where one body's gravity balances another's. Spacecraft can orbit there while burning very little fuel. In 1978, NASA's International Sun-Earth Explorer 3 was the first mission to use low energy orbits around a Lagrange point. Later, using low energy paths between Earth and the Moon, controllers at NASA's Goddard Space Flight Center, Greenbelt, Md., sent the spacecraft to the first encounter with a comet, Comet Giacobini-Zinner, in 1985.

The elaborate flight path for Genesis was designed for the spacecraft to leave Earth and travel to orbit the Lagrange point. After five loops around this Lagrange point, the spacecraft will fall out of orbit without any maneuvers and then pass by Earth to a Lagrange point on the opposite side of the planet. Finally, it will return to Earth's upper atmosphere to drop off its samples of solar wind in the Utah desert.

The problem with all these 'slow-boat' ideas is that for manned flights they increase all of the known stresses to travelers including radiation exposure. These forms of travel may, however, be viable for the transport of cargo or scientific equipment across the solar system where high mass and low cost are desirable goals, and it is not too critical that the payloads take years to get to even the closest destinations.

Medium-speed Travel

This is pretty much what we are doing today, largely with chemical rockets. The actual travel times for past missions are indicated, along with information about the technology.

Mission	Isp	Destination	Duration	Arrival
LRO	311	Moon	5 days	6/23/2009
Venus Express	311	Venus	5 mo	4/11/2006
MRO	311	Mars	7 mo	3/10/2006
Viking 1	444	Mars	10 mo	6/19/1976
Magellan	220	Venus	1.2 yrs	8/1/1990
Pioneer 10	316	Jupiter	1.7 yrs	12/4/1973
Deep Space 1	2000	Borrelly	3.0 yrs	9/22/2001
Voyager 1	444	Jup-Saturn	3.1 yrs	11/12/1980
Dawn	3,100	Vesta	3.9 yrs	7/16/2011
NEAR	266	Eros	4.0 yrs	2/14/2000
Galileo	274	Jupiter	6.1 yrs	12/8/1995
MESSENGER	274	Mercury	6.5 yrs	3/18/11
Cassini	272	Saturn	6.8 yrs	7/1/2004
Dawn	3,100	Ceres	8.6 yrs	3/6/2015
New Horizons	275	Pluto	9.6 yrs	7/14/2015
Rosetta	440	Comet 67P	10.6 yrs	6/8/2014
Voyager 2	444	J-S-U-N	12 yrs	8/25/1989

Dark shaded=ion rocket   Light shaded=gravity assists

The basic strategy is to use one or more Hohmann transfer orbits to travel from Earth to the destination. This works for trips to the moon and Mars because the required velocity changes are small and can be accommodated by chemical rockets. For other destinations beyond the asteroid belt, higher delta-Vs are needed. To avoid carrying a lot of fuel, multiple gravity assists are used to inexpensively boost spacecraft speeds for travels to Venus,

Mercury, Jupiter and beyond. A typical Hohmann Transfer to Mars by the MAVEN spacecraft looks like this.

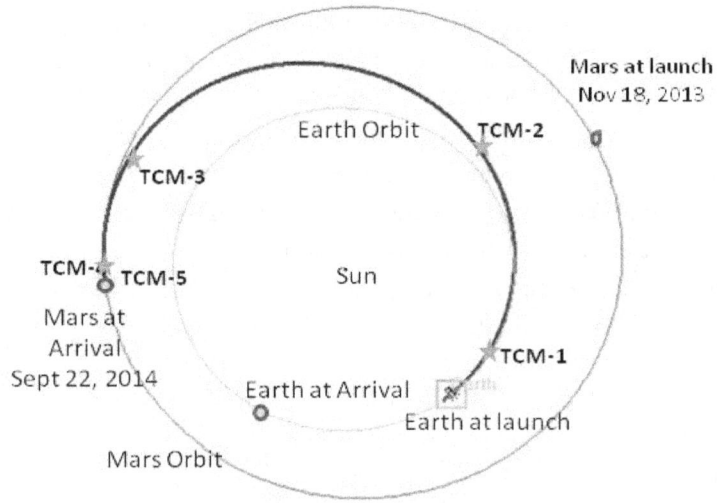

We can also include multiple gravity assists to get still-higher delta-Vs for travels to the outer solar system, or to the inner planets.

The ion rocket missions use very low thrust engines and perform a steadily expanding spiral path to get to the destination, like this trajectory taken by Dawn.

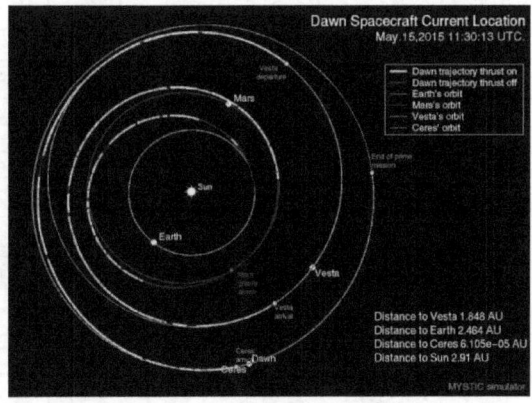

High-speed travel

Although we are currently limited by chemical rockets, gravity assists and low-thrust ion technology, the future will not be. To avoid increased radiation exposure and a variety of other psychological and physical problems, high speed interplanetary travel will be a minimum requirement for journeys into the solar system. Because these hazards, and their mitigation, begin to escalate after six months in space according to long-term studies performed in the International Space Station, it is likely that trip times less than one month will eventually become a goal.

One approach is to give your rocket a huge speed wallop at the start, then coast to the destination and give your spacecraft another huge speed change to slow it down to capture speed. This is basically what we do with current chemical rockets that have low Isps. The difficulty for high acceleration trajectories is that humans can only withstand a limited number of 'Gs' of acceleration

Accel. time (hours)	Max V (k/s)	Mars (days) 78	Jup (days) 629	Sat (days) 1275	Urn (days) 2724	Nep (days) 4351	Plu (days) 4280
1	35	26	206	418	894	1427	1404
2	71	13	103	209	447	714	702
3	106	9	69	139	298	476	468
4	141	7	52	104	224	357	351
5	176	5	42	84	179	286	281
6	212	5	35	70	149	238	234
7	247	4	30	60	128	204	200
8	282	4	26	53	112	179	176
9	318	3	23	47	100	159	156
10	353	3	21	42	90	143	141

If we don't ask for now the difficult question of how we maintain 1-G accelerations for hours at a time, for a comfortable 1-G boost, the table shows the kinds of travel times that are possible for reaching different destinations including the coast phase when the planets are at their closest to Earth's orbit (called opposition). For example: If we accelerate for 10 hours at 9.8 m/sec² our final speed becomes $V = 9.8$ m/sec² x 3600 sec/hr = 353 km/s. The distance traveled from Earth orbit during the acceleration phase is just $d = 1/2aT^2 = (0.5)(0.0098 \text{ km/sec}^2)(10\text{hx}3600\text{s/h})^2 = 6.4$ million km. For Mars at its closest (opposition)the table says its distance is '78' million km, so the time taken is 20 hours + (78 million – 2x6.4 million)/353 = 71 hours or 3 days. The acceleration and deceleration amounts to 20 hours of this trip time. For Pluto, the journey takes 141 days and the 20-hour acceleration/deceleration phase is insignificant.

A second approach is to apply a constant acceleration for half the trip, then a constant deceleration for the second half to slow down to capture speed at the destination. This strategy has been used repeatedly in science fiction.

The story *Tau Zero*, by Poul Anderson, has a spaceship using a constant acceleration drive. The spacecraft of George O. Smith's *Venus Equilateral* stories are all constant acceleration ships. Normal acceleration is 1 G, but in "The External Triangle" it is mentioned that accelerations of up to 5 G are possible if the crew is drugged with gravanol to counteract the effects of the G load. Spacecraft in Joe Haldeman's novel *The Forever War* make extensive use of constant acceleration; they require elaborate safety equipment to keep their occupants alive at high acceleration (up to 25 G), and accelerate at 1 G even when "at rest" to provide humans with a comfortable level of gravity. In the "*Known Space*" Universe constructed by Larry Niven spaceships use constant acceleration drives in the form of a Bussard ramjet for interstellar travel. In "*The Sparrow*," by Mary Doria Russell, interstellar travel is achieved by converting a small asteroid into a constant acceleration spacecraft. Force is applied by ion engines fed with material mined from the asteroid itself. In the novel "*2061: Odyssey Three*" by Arthur C. Clarke, the spaceship Universe, using a 'muon-catalyzed fusion rocket', is capable of constant acceleration at 0.2 G under full thrust. The *Hidden Worlds* spaceships of F.M. Busby's *Rissa Kerguelen Saga* utilizes a constant acceleration drive that can accelerate at 1 G or even a little more.

We can revise the previous table calculations to show this approach with times rounded to the nearest day. The results are impressive, and clearly with this kind of technological approach, the entire solar system becomes just as accessible as commercial tourist vacation tours on Earth today!

Gs	Max V (k/s)	Mars (days)	Jup (days)	Sat (days)	Urn (days)	Nep (days)	Plu (days)
0.2	392	5	13	19	27	34	34
0.4	554	3	9	13	19	24	24
0.6	679	3	8	11	16	20	20
0.8	784	2	7	9	14	17	17
1.0	876	2	6	8	12	15	15
1.2	960	2	5	8	11	14	14
1.4	1037	2	5	7	10	13	13
1.6	1108	2	5	7	10	12	12
1.8	1176	2	4	6	9	11	11
2.0	1239	1	4	6	9	11	11

So, these two tabulated options express two different technological challenges. Can we create a rocket engine that can sustain 1G acceleration for several hours? If so, then the first table represents what we can do at that technological level with a short period of acceleration followed by a long period of coasting at maximum speeds of a few hundred km/sec, followed by a short period of deceleration once we approach our destination. Under these conditions, we can travel to Mars in a week, or Pluto in a few dozen months. In the more advanced technological attainment, we can sustain 1-G accelerations for several days and reach speeds from 500 to 1000 km/sec, making the journeys ten times faster, with Pluto a two-week jaunt. At these timescales, there are no cumulative health effects and radiation exposure is minimal. There is also no time for the crew to get bored or depressed.

To design rocket engines that can produce sustained accelerations of 0.2 to 1.0 Gs, we cannot use conventional chemical rockets because their Isps are far too low, and they operate for only a few minutes. This also makes their payload to fuel ratios (below 10%) very unfavorable, carrying about 10 kg or more of fuel for every 1 kg of payload. Only ion or nuclear rockets can at least theoretically produce constant thrust over long time periods. For example, the NASA Dawn spacecraft used ion engine technology, which operated the engine continuously for hundreds of days. (Image credit:NASA/Dawn)

Nuclear Rocketry

Most people think that Star Trek-style (impulse power) nuclear rockets are a thing of the future, but the fact is we had them in the 1960s. In fact, both the Soviets and US engineers started looking into this 'peaceful use of the atom' technology in the 1950s. They were very impressed by it, and immediately saw its potentials for interplanetary travel.

The Nuclear Engine for Rocket Vehicle Applications (NERVA) program was a joint effort between NASA and the US Atomic Energy Commission. It was managed by the Space Nuclear Propulsion Office, which had begun work on nuclear rockets in 1952. This research accelerated so quickly that by 1961, the NASA Marshall Spaceflight Center started using nuclear rockets in their mission planning, with the first launch to be in 1964 as a final demonstration of the space-worthiness of these engines. The NERVA engine was built by Aerojet and Westinghouse. The first of these actually built and tested in a spaceflight configuration was the Kiwi-B4 engine that produced 70,000 pounds of thrust (330,000 Newtons). The NERVA NRX/EST engine in 1966 ran for two continuous hours. The NERVA-XE engine tests ran for 115 minutes and as a result, it demonstrated that nuclear engines were now flight-ready as a technology. The nuclear engine program had demonstrated thrusts as high as 250,000 pounds, 90 minutes of continuous and controllable thrust delivery, and thermal power equal to 4,500 megawatts.

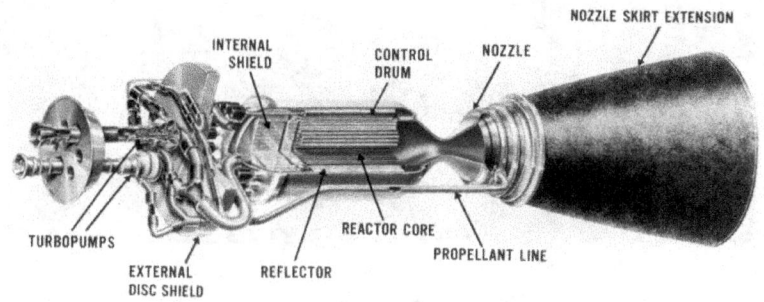

The plan was to use a NERVA engine as the third stage of the Saturn V rocket, and plausibly get to Mars by 1978. There were even plans to use this engine as the work-horse to establish a large lunar colony by 1981. These plans were canceled in 1972 once President Nixon came into office and decided that the Saturn V and the Apollo program were no longer needed to prove US space superiority during the Cold War. Without the Saturn V, there was no way to place the heavy nuclear engine into space, even though once there it would dramatically out-perform any chemical rocket.

Here's a conceptual illustration of a spacecraft for a manned Mars mission proposed by NASA's Wernher von Braun in August 1969. Two spacecraft would make the trip in tandem, each one powered by three NERVA-type engines. (Image credit: NASA)

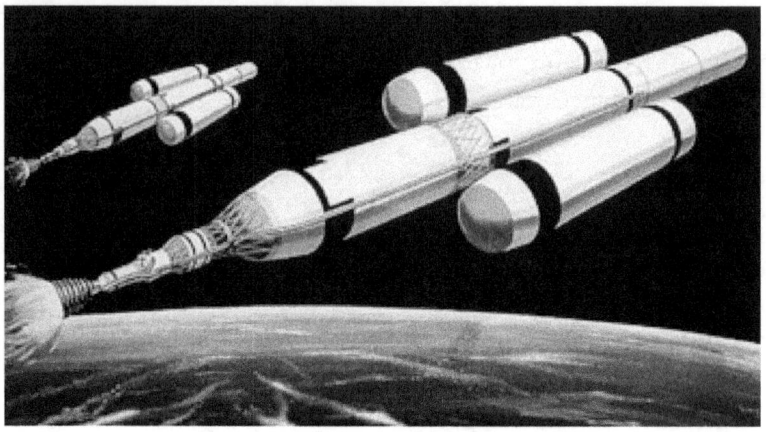

The Space Nuclear Propulsion Office planned to build ten nuclear engine-based vehicles, six for ground tests and four for flight tests, but the development program was delayed after 1966 as NERVA became a political hot-potato in the debate over a Mars mission. The nuclear–enhanced Saturn V would carry two to three times more payload into space than the chemical version, enough to easily loft 340,000 pound space stations and replenish orbital propellant depots. Ultimately, the NERVA program became so closely linked to plans to go to Mars that, when Congress finally balked at the expense and folly of going to Mars, the NERVA program no longer had a 'customer' to serve. The NERVA program was finally terminated altogether on January 5, 1973.

So for decades, the popular story seemed to be that we didn't have nuclear rocket technology and would have to get by with impressive chemical rocket-based systems like the Space Shuttle. But then something amazing happened.

In 2000, NASA proposed *Project Prometheus*. Its goal was to develop nuclear-powered systems for long-duration space missions. Now, this wasn't nuclear engine technology, but the far more practical power plant technology. NASA had for decades been shoehorned into missions that used heavy solar panels or radioisotope thermoelectric generators (RTGs) to generate a few hundred watts to run scientific equipment. Now NASA wanted to leap into multi-kilowatt 'stadium lighting' systems based on small nuclear fission reactors. For the Jupiter Icy Moons Orbiter (JIMO), a spacecraft designed to explore Europa, Ganymede, and Callisto, NASA intended to use the first of what they hoped would be the new generation of mini-reactors flown in space. The reactor would heat a fluid and run a steam turbine to generate electricity. The electricity would then power scientific instruments and an ion-propulsion unit. But this was not to be either. The JIMO mission was cancelled in 2006, and this act also canceled the *Prometheus Project*. (JIMO design: Image credit:NASA/JPL)

NASA chief, Mike Griffin, told a U.S. Senate subcommittee in May 2005 that JIMO was in his opinion, *"too ambitious to be attempted."* Instead, the Juno mission to Jupiter was funded, and this $1.1 billion spacecraft carries huge solar panels to generate its electricity using conservative 'off the shelf' technology.

Radioisotope Nuclear Rocket

One simple design is the radioisotope rocket. For decades, NASA has used radioisotope thermoelectric generators (RTGs) to create

295

electricity for deep space missions operating beyond the orbit of Jupiter where solar panels are impractical and have to be very large and massive. The first RTG launched into space by the United States was SNAP 3 in 1961, aboard the Navy Transit 4A satellite. One of the first terrestrial uses of RTGs was in 1966 by the US Navy at uninhabited Fairway Rock in Alaska. RTGs were used at that site to generate electricity until 1995. RTGs operate by using a source of radioactivity to produce heat, and then using a thermocouple to convert temperature differences directly into electricity.

The mixture of radioisotopes and the heat conductivity of the RTG are balanced so that the maximum temperature does not exceed about 900 kelvins. You don't want the RTG to melt, afterall!

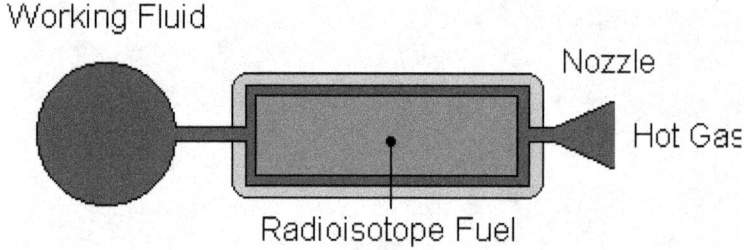

By redesigning an RTG to heat a fuel to 2,000 C, a 14 kg, 5 kilowatt reactor can generate a thrust of about 1.5 N and an Isp of 800 seconds. TRW demonstrated a Polonium-fueled thruster that operated for 65 hours in 1965. This basic idea of heating a propellant by passing it through a material in which fission is taking place, is the basis for a rocket technology called the Nuclear Thermal Rocket (NTR), which has been investigated by NASA for over 60 years. NTRs are the main nuclear rocket designs being investigated today. In fact, the vehicle depicted on the opposite page is the Copernicus: a nuclear thermal rocket upper stage that has been proposed for the Space Launch System and the eventual

2030's mission to Mars. (Image Wikipedia: NASA/Pat Rawlings (SAIC))

Although development of nuclear thermal rockets seemed to have stopped in the 1970s, it was such a seductive next step in rocket development that the research continued through military efforts. Project Timberwind was funded by the Strategic Defense Initiative from 1987 through 1991, and totaled $139 million (in 1991 dollars). It was later transferred to the Air Force Space Nuclear Thermal Propulsion (SNTP) program. Advances in high-temperature metals, computer modeling, and nuclear engineering in general, resulted in dramatically improved performance over the designs created in the 1960s. For example, the NERVA engine was projected to weigh about 6800 kg, the final SNTP design had a mass of only 1650 kg, while further improving the specific impulse to 1000 seconds. The SNTP program was intended to develop upper-stages, which would not operate within the Earth's atmosphere, but the program was terminated in January 1994.

In 2012, NASA and the Russian Federal Space Agency revived NTR technology and coordinating a new $600 million joint engine project along with potential involvement from France, Britain, Germany, China, and Japan.

Interplanetary Travel

Meanwhile, the NASA Marshall Space Flight Center has been developing its own Nuclear Cryogenic Propulsion Stage for the Copernicus upper stage of the Space Launch System. This upper stage would not initiate a fission reaction until safely out of the atmosphere. To test the design and improve it without any test flights, which are banned, Marshall engineers have created a simulated test bed called NTREES. The role of the nuclear reactor in real life is simply to heat the hydrogen fuel flowing through it to temperatures over 2000 C. What NTREES does is to provide a simulated non-nuclear ceramic core that does this work with no nuclear reactions involved. *"The information we gain using this test facility will permit engineers to design rugged, efficient fuel elements and nuclear propulsion systems,"* NASA researcher and Manager of the NTREES facility, Bill Emrich, said. *"It's our hope that it will enable us to develop a reliable, cost-effective nuclear rocket engine in the not-too-distant future."*

At the present rate of progress and funding, and with a manned journey to Mars pegged at some time in 2033-35, the NTR design ws until very recently a part of the established planning (Called DRA 5.0) for the mission. Chemical rockets will not be able to transport humans to Mars quickly enough so that they do not absorb dangerous levels of radiation from the much-longer trip.

The NTR engine will power the Copernicus module using a standard uranium-235 nuclear reactor, which will heat the hydrogen gas to provide thrust. Three of these engines would be sufficient to produce a total of 75,000 pounds of thrust. The main engine core is an hexagonal graphite cylinder containing U-235 microspheres. Coolant channels run down the axis of the core, and are lined with zinc carbide to prevent erosion of the graphite from the high temperature hydrogen flow. The U-235 heats the liquid hydrogen to temperatures of 2600 k, and at a flow rate of 12.6 kg/sec, the engine has an Isp of 900 seconds (DRA 5.0).

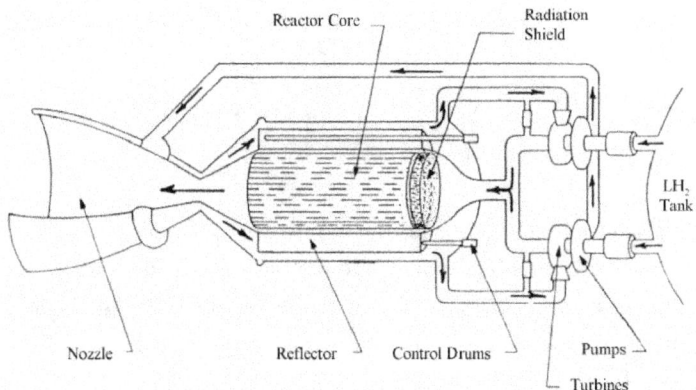

The three engines would be loaded with a total of 120 kgs of enriched Uranium-235, less than 1% of which would be fissioned during a round trip mission to Mars, so they can be recycled by just refueling with more hydrogen propellant. The spacecraft would make the 40 million mile trip to Mars within 100 days. Unofficial cost estimates of about $40 billion for a nuclear-powered manned Mars mission seem reasonable when spread over a multi-year timeframe. If we started in 2015, the annual cost would be about $2 billion, which unfortunately cannot be accommodated by the current NASA budget unless new money is added, or the ISS is decommissioned, which are very unlikely.

The design of an NTR is in many ways quite simple. The reactor produces a reliable and constant heat source that heats a propellant to a high velocity to produce the thrust. A relatively simple formula relates the relevant physical parameters needed to achieve a given exhaust speed.

$$c = \sqrt{\frac{2\gamma}{\gamma-1}\frac{R_u}{M}T_c}$$

The highest exhaust speeds, c, occur for propellants with the lowest mass, M, which is why hydrogen is often used. With R being the universal gas constant with a value of 8320 J/(kg mole K), molecular hydrogen gas has gamma=1.4 and a mass M = 2 AMU/mole, so

$$c = g \times Isp = \sqrt{\frac{2.8 \times 8320 T}{0.4 \times 2}} = 172\, T^{1/2} \quad \text{meters/sec}$$

If the molecular hydrogen propellant is heated to T =2500 K, then the exhaust speed is just 8.6 km/sec and the Isp is 877 sec. When core elements are cladded with zinc carbide, temperatures to 2700 K can be reached for Isp = 911 sec. The main limitation for the thrust of a NTR is the temperature of the exhaust. For Isps near 1200 sec, the temperature for a hydrogen propellant has to be above 4000 K. This will melt or vaporize just about any material without considerable active cooling.

By the way, although we call the core of a nuclear engine a fission reactor, it really doesn't work anything like the kinds of reactors we are normally familiar with. For power generation, we build nuclear reactors that sit in a containment vessel and basically boil water to steam to run an electrical steam turbine. The reactor is carefully regulated using control rods to make sure it never 'goes super critical', which means that the core never heats up over 300 C. Newer designs called High Temperature Gas Cooled Reactors are being developed that operate at temperatures up to 800 C. For nuclear rockets, the goal is to get the reactor to operate at many thousands of degrees and nearly at its melting point! You are literally trying to manage something almost as hot as the surface of our sun and keep it operating for months or years to produce a well-regulated thrust.

The gas temperature that can be reached in a solid-core nuclear engine is limited by the melting point of the core's materials. But fission reactions can also take place in molten or gaseous cores. Nominal core temperatures for solid, molten, and gaseous cores are 2,750 K (4,490°F), 5,250 K (8,990°F), and 21,000 K (37,340°F), respectively. The much higher temperatures of molten and gaseous cores would produce higher exhaust velocities than are possible with solid cores. These would produce exhaust velocities of 9, 12, and 25 kilometers/sec, and Isp's as high as 7,900 sec.

In NASA's Design Reference Architecture (DRA) 5.0.1 for a manned mission to Mars in the 2030s, NTP is the preferred propulsion technology because it has heritage and higher performance than current propulsion systems available at the thrust levels necessary for manned missions. The Nuclear Cryogenic Propulsion Stage (NCPS) project is engaged in conceptual design of a small 33,400 Newtons, and a full-size 111,200 Newton-class nuclear rocket. The small engine could be used for a flight demonstration mission as well as possible robotic probe missions. The large nuclear engine is intended to be used in a clustered configuration for human exploration missions.

Currently the highest performance propellant combination commonly used for chemical systems is hydrogen and oxygen. At a mixture ratio of about 6 to 1, the combustion temperature is approximately 3420 K, the average molecular weight is 13.8 g/mol, and the maximum theoretical specific impulse is near 480 seconds. In a nuclear rocket, the propellant is typically H_2 with a molecular weight of 2.016 g/mol and is heated to temperatures near 2700 K. Based on the equation for specific impulse, the nuclear rocket has twice the expected specific impulse, approximately 900 seconds, as hydrogen and oxygen chemical propulsion.

In 2012, the National Research Council committee reviewing the NASA Technology Roadmaps recommended 16 technologies out of the 330 reviewed for NASA to pursue as high priority—the NTR was third. Consequently, in 2012, NASA initiated the Nuclear Cryogenic Stage Program (NCSP) to start the development of a NTR. The NCSP is examining 1) fuel materials, 2) engine design, and 3) laboratory testing of fuel elements in a non-radiation environment, and 4) ground testing of the NTR at full power for full duration. Engineer Michael Houts is leading the nuclear-propulsion project with a budget of US $3 million — minuscule in comparison with the $1.3 billion that NASA will spend on space-technology research and development in the 2012 fiscal year. *"The funding at times has gone to zero,"* says Houts. *"You lose the teams and the momentum."* But at least there is a formal recognition that nuclear engines must play a key role in any manned travel to Mars and beyond.

Mason Peck, NASA's chief technologist, says that he will use the NRC priority list as a guide when setting funding in future, but developing fission power for space will require not only money, but also political will. The image of a nuclear-powered spacecraft blowing up on the launch pad or on its way to orbit is a powerful deterrent. The risk of nuclear material contaminating Earth after an accident is negligible because in the current design the NTR reactor would not be started until the system was in orbit.

The full power ground testing issue is considered one of the trickiest issues in developing the Mars Mission NTR. It cannot be tested in space due to the enormous expenses, so some kind of ground testing facility has to be created that will not scare Congress or the residents of the state in which the facility is created. Luckily with the nuclear programs of the 1960s we have been here before.

One promising strategy is to use the facilities at the Nevada Test Site used for nuclear weapons testing by injecting the exhaust from the NTR directly into the sub-strata at the Nevada Test Site. The 'SAFE' concept as it is called, would be simpler, less expensive to operate, and allow any power level of NTR to be tested for full duration. The estimated cost of testing the NTR is around $45 million using the SAFE method.

Another approach, perhaps more well-suited to testing other NTR concepts, is currently housed in the Marshall Center's Propulsion Research and Development Laboratory. Nuclear thermal research at the Marshall Center is part of NASA's Advanced Exploration Systems (AES) Division, managed by the Human Exploration and Operations Mission Directorate and includes participation by the U.S. Department of Energy. The test facility called "NTREES," short for the Nuclear Thermal Rocket Element Environmental Simulator, is already licensed by the Nuclear Regulatory Commission, and is certified to test prototypical nuclear rocket fuel elements. These are identical to the fuel elements used in an NTR, but because the test facility uses non-nuclear heating instead of nuclear fission, the fuel does not become radioactive during the test and can be easily handled and examined once the test is complete.

So, rather than being just some theoretical toy, at least NTR engines have been fully resurrected since the 1960s and are now fully integrated into the missions to Mars that are hopefully to occur in the 2030s. Although the funding for the continued research and development of these engines has been a roller coaster, many millions of dollars have already been spent on making newer generations of designs flight-ready. But the major challenge is dealing with the perception by most people that anything 'nuclear' is dangerous and will result in either a lethal 'dirty bomb' or a replay of Hiroshima. Although Congress is interested in the idea of going to Mars, they as yet have not fully understood the implication that such trips can only be safely accomplished by greatly reducing radiation exposure and hence travel times for the astronauts. This will only be possible with a NTR-based system.

The bottom line for nuclear engines is that, the higher you can make the exhaust temperature, the higher will be the exhaust velocity and Isp. All the reactor does is heat the fuel as high as we can get it, but there are other ways to heat fuel to even higher temperatures. NTR engines are pretty simple in design, but they do not exhaust the possibilities for nuclear rocket design and fuel heating. Engineers consider what are called Open Cycle designs, where the nuclear reactor ejects fuel often mixed with the radioactive byproducts of the fission reactions. There are also Closed Cycle systems where the reactor is separate from the fuel and no radioactive materials exit the engine.

Gas Core Nuclear rockets

NTR are Open Cycle systems, which is a class that also includes Gas Core Nuclear rockets (GCNRs), where the fissioning material is mixed into the fuel itself. The reactor core may be either a gas or plasma. GCNRs may be capable of creating specific impulses of 3,000–5,000 sec and exhaust velocities up to 50 km/s. Because the propellant is mixed with the reactor material, the energy from the fissioning process heats the fuel by absorbing the ultraviolet radiation given off by the fissioning gas, at a working temperature of around 25,000 °C.

The problem is that hydrogen gas is almost transparent to ultraviolet light, so to work around this problem, the propellant is seeded with an opaque solid or liquid such as tungsten or tantalum hafnium carbide (Ta_4HfC_5) soot particles, which melt near 4,200 °C (7,600 F), or liquid tungsten droplets (boiling point 5,550 °C). These particles would make up 4% of mass of exhaust gas, which considerably increases propellant cost while slightly lowering the rocket's specific impulse.

If you want to get specific impulses of 5000-7000 s to make really fast interplanetary travel possible, you will need exhaust gas temperatures of 50,000-100,000 Kelvin, at which point no solid or liquid material would survive, and all propellants would become transparent. Another important aspect to GCNRs is the impact of the rocket acceleration on the containment of the fuel in the fuel bubble. Some computer simulations show that a rocket acceleration of only 0.001 g's will cause buoyancy effects to decrease core containment by 35% if all other flow-rates are held constant from a zero-g startup.

Currently, research into the GCNR concept is underway at the Los Alamos National Laboratory under a program from the NASA Marshall Space Flight Center. Although earlier designs from the 1960's proposed a spherical containment vessel, modern designs prefer tokamak-style toroidal systems.(Image credit: NASA/Bryan Palaszewski)

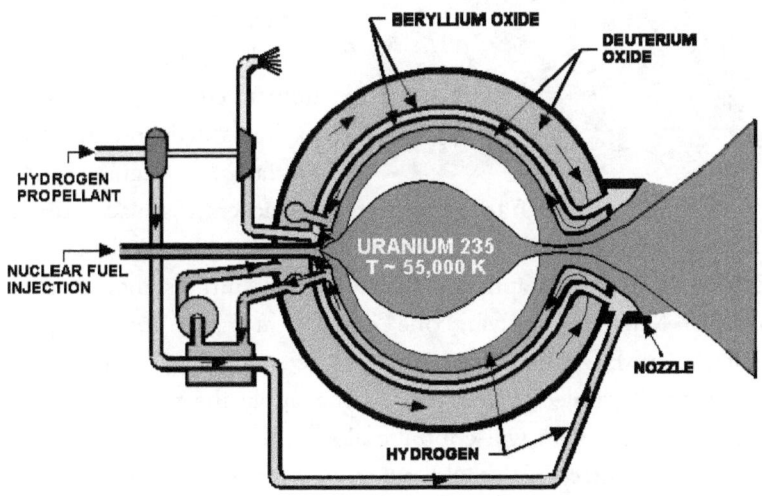

Interplanetary Travel

The research into GCNRs was only funded for a short time, but the infusion of money into a cash-strapped research area at least for a short time led to a flurry of activity in designing physics-based models that have greatly improved our technical understanding of how GCNRs work. Eventually, the hope is to be able to fully examine critical issues such as shear-flow-turbulence losses of the uranium, mixing caused by displacement of the vortex due to acceleration, the need for sufficient residence time of the propellant in the chamber, fission product removal, and stability of the vortex. If these high-Isp engines can be developed and perfected, they hold enormous promise for making interplanetary travel a far faster undertaking than even the NTR designs now being considered for the first manned trips to Mars.

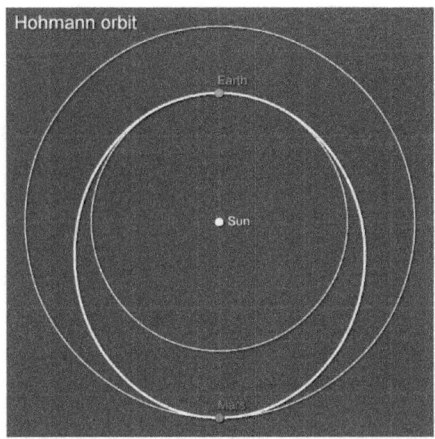

Missions to Mars can be thought of as basically two distinct types. The first type relies on low speed transits with chemical rocket-powered spacecraft placed in Hohmann Transfer orbits. This is what we have been doing for all the unmanned scientific payloads delivered to Mars so far. The figure shows the circular orbits of Earth and Mars. A spacecraft is placed on an elliptical orbit such that its perihelion (closest to sun) distance is Earth's orbit, and its aphelion (farthest from sun) distance is at the orbit of Mars. By applying one 'delta-V' at Earth and a second 'delta-V' at Mars you can coast the entire way from Earth to Mars by taking this trajectory. If you do not apply the second velocity change at Mars, you will continue along the orbit to location you left, but of course Earth will have moved far from this now-empty orbital position. For manned flights, this approach will still be used though with slightly shorter transit times than 9 months one-way.

The second type of mission has not been tried yet. It consists of much faster transits, higher delta-V breaking requirements at the target planet, and far shorter stay times on Mars, roughly 30 to 90 days. The difficulty is that chemical rockets are too slow and massive for this faster method to be used currently.

With the GCNR, a third type of mission can be considered -- the point-and-shoot, which is very popular in science fiction literature: Just point your rocket at the place you want to end up, and start the rocket engine! For this type of travel, the ship transits to Mars in a few months, stays from 30 to 60 days, and returns to Earth in a few months. This mission requires very high delta-V burns at all four staging points - Trans-Mars Injection (TMI), Mars Orbital Insertion(MOI), Trans-Earth Injection(TEI), and Earth Orbital Insertion(EOI). In order to be able to execute such a mission with a reasonable mass fraction of the ship in orbit, the propulsion system must have a specific impulse of around 2000 seconds or higher.

According to the Los Alamos National Laboratory report 'Reducing the Risk to Mars: The Gas Core Nuclear Rocket' by Howe, DeVolder, Thode and Zerkle, the delta-Vs for a fast transit mission occurring in the Year 2011 are 6.4, 12.3, 15.3, and 14.7 km/s for the four burns at TMI, MOI, TEI, and EOI respectively. Thus, the total delta-V for all four burns is near 50 km/s. Compare this with the typical total delta-V of about 6 km/s current spacecraft use to get to Mars via Hohmann Transfer. If the GCNR has a specific impulse of 3000 seconds, then just under 20% of the total ship mass in LEO will be payload and structure-- the rest will be fuel. The mission will require 4 kg of fuel for every kg of payload to perform the entire mission. Alternatively, for a solid core nuclear rocket to achieve these delta-Vs and perform this mission would require over 100 kg-fuel per kg-payload, a mass fraction in LEO of less than 1%.

Closed Cycle Engines.

All of the previous designs are very messy. They eject heated propellant through fission reactors, and in some cases mix highly radioactive debris with the exhaust. This is because the propellant is in direct contact with the fissioning material that is heating it. Another approach is to keep the propellant physically isolated from the reactor core so that what comes out the rocket is completely harmless, well, except for its insanely hot temperatures! These systems are called Closed Cycle Engines.

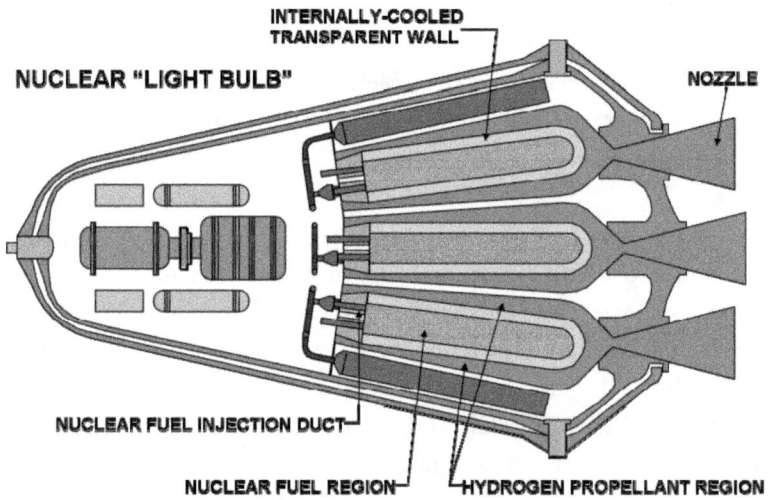

A 'nuclear lightbulb' engine is a hypothetical type of gas core reactor rocket that separates the nuclear fuel from the coolant/propellant with a quartz wall. It would be operated at such high temperature (about 25,000°C) that the vast majority of the electromagnetic emissions would be in the hard ultraviolet range.

Fused silica is almost completely transparent to this light, so it would be used to contain the uranium hexafluoride reactor core, and allow the light to pass through the quartz walls and heat reaction mass in a rocket or to generate electricity using a heat engine or photovoltaics.

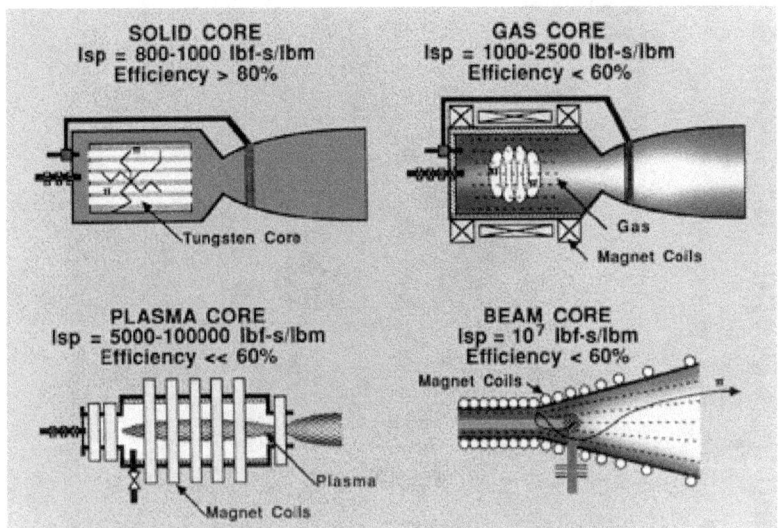

Beam Core: In a nuclear photonic rocket, a nuclear reactor would generate such high temperatures that the blackbody radiation from the reactor would itself provide significant thrust. Photonic rockets are technologically feasible, but rather impractical with current technology.

Nuclear rockets are all based on the idea of heating a low-weight gas (hydrogen) to the highest possible temperatures, and in high enough quantities, to generate thrust. Another strategy is to accelerate the heaviest materials to the highest speeds and eject them. This is the strategy used in ion engines, which we will discuss next.

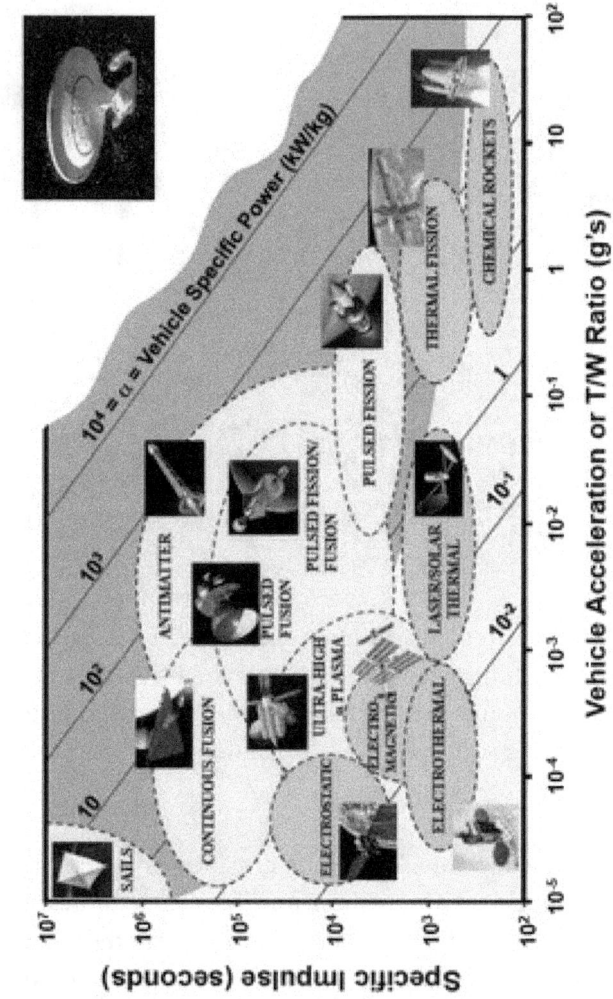

Ion Engines

Ion engines are physically very simple, and use a propellant, usually a heavy element like xenon that has a low ionization energy. This stream of gas is fed to an ionization chamber, and then accelerated towards a porous electrical grid with a voltage of several thousand volts. The ions reach a speed of a few kilometers/sec as they pass through the acceleration grid, and then electrons are added back into the exhaust flow to make the flow electrically neutral, otherwise the exhaust flow will eventually turn around and clobber the spacecraft. Typically more than one of these grids are used to control the flow and acceleration of the ions. That is why these

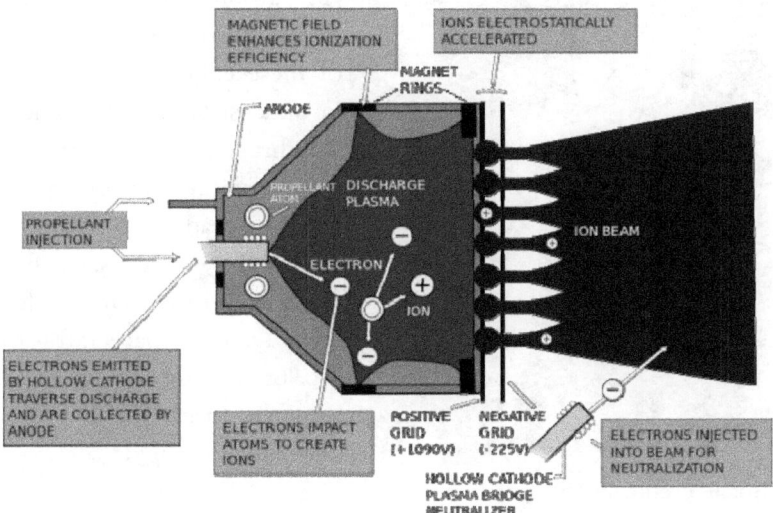

311

designs are called gridded ion engines.

If xenon ions are to be accelerated to a velocity of 40 km/s, which is appropriate for a value of Isp of about 3500 seconds, a potential of 1000 volts is needed. This is the potential of the inner or 'screen' grid and of the thruster body. The next grid, the accelerator, is at a negative potential of about -250 volts, to provide focusing of the ions and to enable the required current to be extracted. Deceleration of the ion flow then follows, via a low negative voltage applied to the outer third grid.

Several different thrusters in this category have flown on a variety of spacecraft. Including experimental devices with grid diameters up to 65 cm, with thrusts up to nearly 1 Newton. Power consumption ranges from a few tens of watts up to more than 10 kW. Specific impulses in the range 2500 to 5000 s have already been achieved, though the actual engine thrusts are very low.

Three NASA missions, SERT-1, Dawn and Deep Space 1, used ion engines to get around in the solar system. Although the thrust of these engines is measured in ounces not in kilopounds, the thrust can be maintained for weeks, months and years, which steadily builds up enormous speeds. For example, it took four days for the Dawn spacecraft to accelerate from 0 to 60 mph using a thrust of just 0.3 ounces. (Ion engine of DS-1.

Image credit: NASA)

SERT-1 and II launched in 1964 and 1970, verified ion thruster technology in space and ran for up to 3700 hours. This technology was then used on the Deep Space-1 mission (1998-2001) that traveled 260 million miles with an engine that ran continuously for over 16,000 hours and was restarted over 200 times. It used 178 pounds of xenon, which was ejected from the engine at speeds up to 88,000 mph. The total speed change using the ion engine was about 4.3 km/sec. The engine used 2100 watts of solar-electric power to create an electrical potential of about 1300 volts to accelerate the xenon ions. It successfully flew past Comet Borrelly and asteroid 9969 Braille. The Dawn mission, launched in 2007 has now visited asteroid Vesta (2011) and dwarf planet Ceres (2015) using a 10 kilowatt ion engine system with 937 pounds of xenon, and achieved a record-breaking speed change of 10 kilometers/sec, some 2.5 times greater than the Deep Space-1 spacecraft.

The NASA NEXT program in 2010 demonstrated over 5.5 years of continuous operation for a 7 kilowatt ion engine. It used 1,900 pounds (862 kg) of xenon and produced a thrust of 0.8 pounds (3.5 Newtons). (NEXT engine test. Credit: NASA)

NASA is not the only space agency that has become familiar with ion engine technology. ESA's SMART-1 technology demonstrator mission to the moon used a Hall-effect thruster (HET) to transfer the spacecraft from a geostationary transfer orbit (GTO) to lunar orbit during November 2004. Japan's Muses-C mission to return a sample from asteroid 1998 SF36 was launched in 2003 and achieved a velocity change of about 3.7 km/s.

The bottom line for ion propulsion is the total electrical power that is available to accelerate the propellant ions. Solar panels are too heavy for the power they produce (60 watts/kg near Earth orbit, but only 3 watts/kg near Jupiter) and do not function well for systems much beyond the orbit of Mars. That means the future of ion propulsion at the levels needed to be most effective, require some kind of fission reactor system (500 watts/kg) to produce the electricity. The history of using reactors in space though trivial from an engineering standpoint, is a politically complex one because of the prevailing fear that a launch mishap will result in a dirty bomb or even a Hiroshima-like event in the minds of the general public and Congress. Various programs have been commenced in several countries to develop suitable reactors for space us, but only those of the Soviet Union have reached operational status.

Early in 1992, the idea of purchasing a Russian-designed and fabricated space reactor power system and integrating it with a US designed satellite went from fiction to reality with the purchase of the first two Topaz II reactors by the Strategic Defense Initiative Organization (now the Ballistic Missile Defense Organization (BMDO). SDIO also requested that the Applied Physics Laboratory in Laurel, MD propose a mission and design a satellite in which the Topaz II could be used as the power source. The outcome of these two activities was the design of the Nuclear Electric Propulsion Space Test Program (NEPSTP) satellite which combines a modified Russian Topaz II power system with a US designed satellite to achieve a specified mission. Due to funding reduction within the SDIO, the Topaz II flight program was postponed indefinitely at the end of Fiscal Year 1993.

Similarly, cancellation was the eventual fate of the US SP-100 reactor program. This program was started in 1983 by NASA, the US Department of Energy and other agencies. It developed a reactor with heat pipes transporting the heat to thermionic converters.

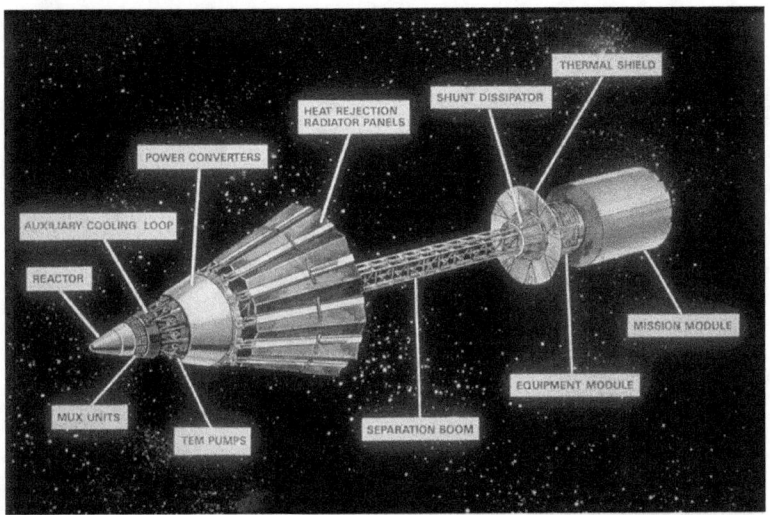

Proposed SP-100 reactor (Image credit: NASA/DoE/DARPA)

Nevertheless, researchers have studied whether ion engine thruster technology operating at a few kilowatts can be scaled up to megawatt systems powered by future reactors in space if we are ever allowed to use them. The results are very promising. Apparently from theoretical and engineering considerations, there is nothing to prevent Isps as high as 5000 from being reached. The figure on the next page shows the kinds of performances that have been predicted for these kinds of ion engine systems for a hypothetical trip to Pluto with an initial mass at LEO orbit given by M0 and for different payload masses Ms. The equivalent ion beam accelerating voltage, V_B, is also shown for xenon propellant.

For only a modest voltage of 3,000 volts and an Isp of 5500, you can deliver a payload of 3.5 tons to Pluto starting from a total rocket mass of 7 tons at LEO! When you get there, you can then run the ion engine as a magnetohydrodynamic plasma generator to

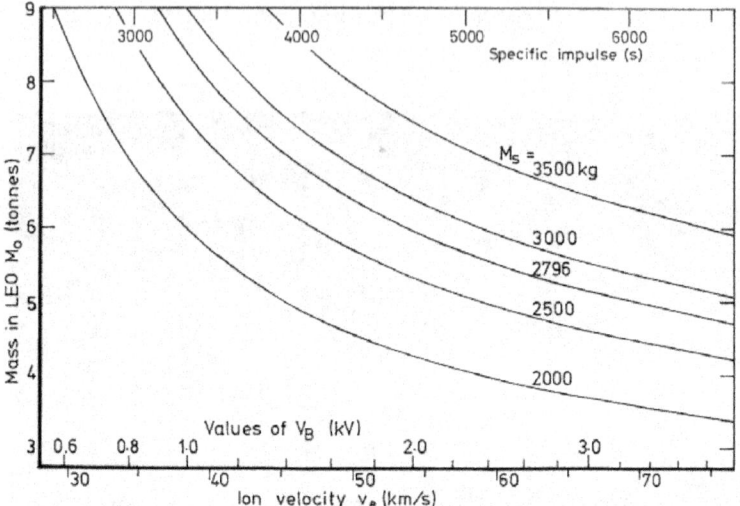

create electricity rather than thrust. If you ask any planetary scientist, there is a lot of really great instrumentation you can pack into a 3-ton spacecraft operating at many kilowatts of power! Now let's think even bigger!

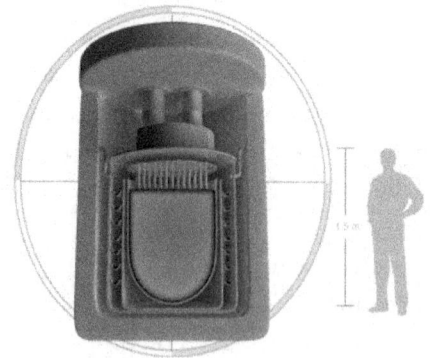

Believe it or not, small nuclear fission reactors are becoming very popular as portable 'batteries' for running remote communities of up to 70,000 people. The Hyperion Hydride Reactor is not much larger than a hot tub, is totally sealed and self-operating, has no moving parts and, beyond refueling, requires no maintenance of any sort. The reactor will output enough to power a community of 20,000 homes, according to the Hyperion Energy Company. Known as the Gen4, it has a mass of about 100-tons and is designed to deliver 70 megawatts of heat (25 megawatts of electricity) for a 10-year lifetime, without refueling.

Of course you cannot just slap one of these onto a rocket ship to provide the electricity for the ion engines, but this technology already proves that fission reactors can be made very small and deliver quite the electrical wallop. The problem for ion rockets is that you have to keep the mass of the power generator small compared to the kilowatts of electricity you generate. Let's look at some popular sources in terms of their 'specific power' in electrical kilowatts per kilogram!

System	Type	Mass (kg)	Power (W)	P/M (W/kg)
Cassini	RTG	68	300	4
Intl. Space Station	Solar	8700	260,000	30
Deep Space-1	Solar	52	2500	48
Geosync. Satellite	Solar	200	16,000	80
Thin film III-V	Solar			100
Hyperion Gen4	Fission	100,000	25,000,000	250
Thin-film SRS	Solar			4300

Meanwhile, thin-film solar arrays produced by SRS Technologies in Huntsville, Alabama have the highest power/mass ratio on record - 4300 W/kg! These arrays can be stowed in a rolled or folded configuration in the launch vehicle and deployed in space by simple boom extension or roller mechanisms. To get a 25 megawatt system it will have a mass of about six tons not including all of the electronics and mechanical support.

Given the Congressional aversion to developing fission-based power for space application, solar power is the most likely technology for ion engines for the foreseeable future. This also limits us to operations in the inner solar system where sunlight is plentiful.

A 6 kW Hall thruster in operation at the NASA Jet Propulsion Laboratory. (Image credit: NASA/JPL)

The Hall-Effect Thruster

The Hall-effect thruster (HET) accelerates the ions entirely within the discharge plasma. The typical HET consists of an annular discharge chamber made from ceramic material, with a ring-shaped hollow anode situated at its closed end. The propellant gas, again usually xenon, is injected through this anode, and the correct application of a magnetic field is critical to success. Lifetime is limited to about 6000 to 8000 hours, but this is adequate for many missions of interest.

HETs with nominal diameters of 5, 7 and 10 cm and thrusts of up to more than 0.08 N have been flown on more than 60 occasions in Soviet and Russian missions. Much larger devices are under development in both Russia and the USA, with thrusts that exceed 1 N.

The Keldysh Research Center's X-85M thruster is designed to operate at up to 1 kV. At 750 Volts its Isp is reported to be 3200 s, which is itself a considerable achievement. An Isp of 4700 s has been obtained at NASA Glenn using krypton in the new 400M thruster. The power input has been steadily increased to over 72,000 watts for an Isp of over 3200 seconds, and a thrust of nearly 3.0 N.

Hall Effect engines have now flown on numerous Russian and Soviet spacecraft with considerable success, and have now been adopted by many European and American organizations, primarily for application to communications satellites. This technology has been used to transfer the SMART-1 spacecraft from its initial GTO to lunar orbit. Numerous mission analyses suggest that this technology is, however, inadequate for the majority of deep space missions.

VASIMR Engines

Another design that will soon be tested on the International Space Station is the variable specific impulse magneto-plasma rocket (VASIMR). In this device, radio waves heat hydrogen gas to extreme temperature to create plasma. Magnetic fields contain the plasma within the engine and direct it out the nozzle to produce low levels of thrust.

NASA was working on one of these engines until 2005. Its development has now been turned over to a private company—*Ad Astra Rocket Co* whose VX-200 rocket prototype operating at 200 kilowatts in 2010 demonstrated its highest power efficiency and performance at the company's Houston laboratory. It achieved a thrust of 5.7 Newtons and an exhaust speed of 50 km/s.

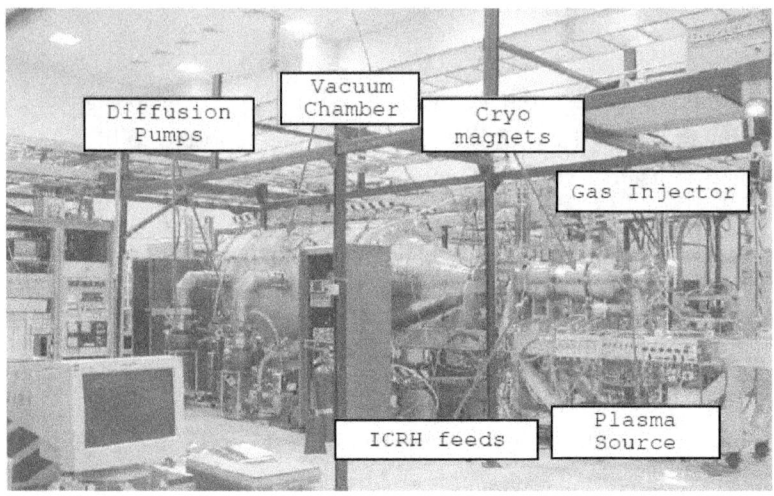

A prototype VASMR engine called the VX-10 was tested at the Johnson Spaceflight Center in Houston…(Sakai, 2004)

A 2004 doctoral thesis by Tadashi Sakai at the Georgia Institute of Technology concluded that a VASIMR-type engine could cut the transit time to Mars from nine months to as little as one month. The caveat is that tens of megawatts of power have to be used, and these kinds of flight-ready power plants are not available.

On December 8, 2008, *Ad Astra* signed an agreement with NASA to arrange the placement and testing of a flight version of the VF-200 on the International Space Station in 2016. The test on the ISS may lead to a capability of maintaining the ISS or a similar space station in a stable orbit at 1/20th of the $210 million per year present estimated cost. The VF-200 flight-rated thruster consists of two 100 kW units with opposite magnetic dipoles so that no net rotational torque is applied to the ISS when the thruster magnets are firing. The VF-200-1 is the first flight unit and will be tested in space attached to the ISS. (Image credit: NASA)

The claims by *Ad Astra* that their VASIMR engines will make the manned trip to Mars in only 40 days has been called into question by NASA engineer Robert Zubrin.

"The Mars trip would require a nuclear reactor system with a power of 200,000 kilowatts and a power-to-mass ratio of 1,000 watts per kilogram. In fact, the largest space nuclear reactor ever built, the Soviet Topaz, had a power of 10 kilowatts and a power-to-mass ratio of 10 watts per kilogram...If generous but potentially realistic numbers are assumed (50 watts per kilogram), Chang Diaz's hypothetical 200,000-kilowatt nuclear electric spaceship would have a launch mass of 7,700 metric tons, including 4,000 tons of very expensive and very radioactive high-technology reactor system hardware requiring maintenance support from a virtual parallel universe of futuristic orbital infrastructure. Yet it would still get to Mars no quicker than the 6-month transit executed by the Mars Odyssey spacecraft using chemical propulsion in 2001, and which could be readily accomplished by a human crew launched directly to Mars by a heavy-lift booster no more advanced than the

(140-ton-to-orbit) Saturn 5 employed to send astronauts to the Moon in the 1960s."

Development of high-thrust ion engines

SERT II was the first NASA-built ion propulsion spacecraft launched in February 1970. This mission included two 15 cm diameter ion thrusters developed by the NASA Lewis Research Center in the 1960s. Using mercury propellant, these thrusters each provided a nominal thrust of 0.029N with an Isp of 4770 s. The mission continued operation until all propellant was exhausted about 11 years from the launch. Since then, many other large aperture ion thrusters have been laboratory-tested and even flown on interplanetary spacecraft. The table shows the properties of some of these systems larger than 20 centimeters. The shaded systems are larger than 30 centimeters.

Name	Year	Thrust mNt	Isp	Input watts
RIT-XT	2003	200	6420	8000
T6 Kaufman	1997	200	5000	
ESA-XX	1991	200	5400	8500
UK-25	1990	320	5000	10000
NSTAR-MESC	1999	92.6		2300
RIT-35	1979	104	5000	7000
BBM-2 MESC	1999	180	3500	4000
TMIT-MESC	2000	153	3710	3500
NEXT-MESC	2003	237	4400	6900
NEXIS-MESC	2004	517	8700	27000
HiPEP MESC	2004	670	9620	40000

For the more challenging interplanetary missions, probably using nuclear power sources, NASA has decided to invest in the development of ion thrusters much larger than NEXT. The more conventional of these is the NEXIS device, where this acronym is derived from "nuclear electric xenon ion system". This is a 65 cm (26-inch) diameter thruster similar to the NSTAR engine that was flown on the Deep Space 1 spacecraft. The initial mission for which it was intended was the Jupiter Icy Moons Orbiter (JIMO), which was canceled in 2005.

The NEXIS ion engine development was being led by JPL, and the objectives included a power consumption of 20kW, an Isp of 7500 s using xenon propellant, with a beam accelerating potential of 8 kV. This photograph was taken when the power consumption was 27 kW, the total ion extraction potential 7 kV, with a thrust of 0.517 Newtons, and an Isp of 8700 s.

At the beginning of their electric propulsion program in 1967 ,engineers at the NASA Lewis Spaceflight Center investigated the possibility of building a 1.5-meter diameter gridded ion engine. Although a prototype was constructed, it never performed up to its expectations for thrust and Isp so the program was abandoned. The difficulties associated with managing the ion flow through such large grids was still present when ion engines 50 cm diameter were being developed in the 1980s. This lead to the conclusion that the 40 cm of the NEXT engine might be the maximum possible with present technology.

From the point of view of many missions, the most important parameter is the exhaust speed of the propellant, v, which determines the value of Isp achieved. This is given by $1/2mv^2 = eV$ so

$$v(\text{exhaust}) = \sqrt{\frac{2eV}{m}}$$

where e is the charge on an electron, m is the ion mass, and the ions are assumed to be singly charged. Thus v is determined by the electrical potential V and by m. In practice, V is the potential of the body of the thruster and of the screen grid, which is typically 1 to 2.5 kV in present designs. As an example of what can be achieved, the values of v and Isp attained using a relatively low atomic mass propellant, argon, are plotted for potentials of up to 40 kV. The latter voltage gives a velocity of over 400 km/s and an Isp of about 40,000 seconds.

The predicted performance of a 30-cm beam diameter thruster using the slightly heavier xenon propellant, a 4-grid system and beam potentials of up to 70 kV give a maximum thrust of 20 Newtons for an input power of 6 megawatts

Among the largest ion engines that have been tested so far is a 5 megawatt engine with a size of 40 cm x 18 cm developed in 1984 by the Culham Laboratory. With a beam energy of 80 kV, the Isp is exceptionally large at 400,000 seconds, and the thrust was 2.4 N.

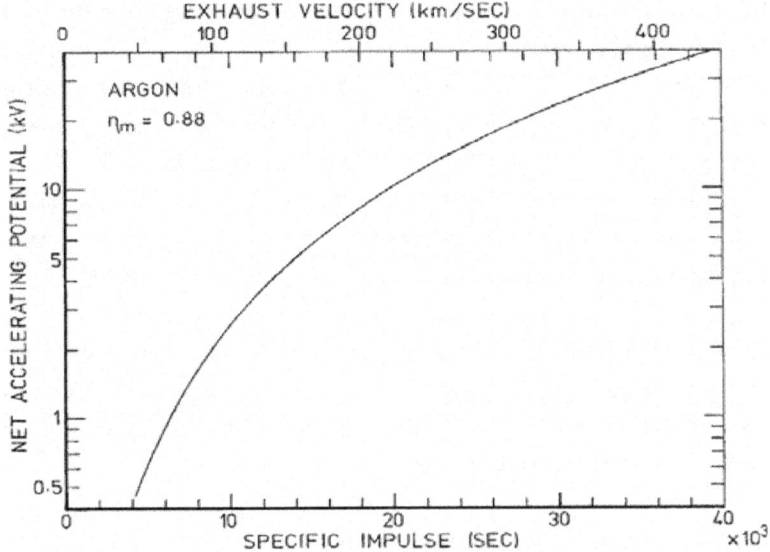

An extensive research study on the design of megawatt ion engines by David Fearn presented at the Space Power Symposium of the 56th International Astronautical Congress in 2005 gave some typical characteristics for engines at this power level. The conclusion was that these kinds of ion engines pose no particular design challenges and can achieve Isps that exceed 10,000 seconds.

	500 kW	1500 kW	1500 kW
Propellant	Xe	Xe	Xe
Grid dia	40 cm	40 cm	40 cm
Prop effi	90%	90%	90%
Isp	8600	15,000	12,000
Thrust	10 nt	18 Nt	22 Nt
Beam Cur	80 Amps	82 Amps	123 Amps
Prop flow	0.12 g/s	0.12 g/s	0.19 g/s
Pwr/Thrust	49 W/mN	83 W/mN	68 W/mN
Thruster mass	32 kg	35 kg	35 kg

Interplanetary Travel

"It is thus concluded that well understood thruster technology, when combined with the 4-grid configuration based on that utilised in CTR ion injection machines, will permit MW power levels to be achieved. Thus a relatively small array of thrusters, with beam diameters not exceeding 40 cm, will be able to consume many MWs, although the SI, using Xe propellant, is likely to be somewhat above 10,000 s. If higher values of Isp are required, the utilisation of lower atomic mass propellants will permit this to be achieved, with an ultimate limit using hydrogen compounds of about 150,000 s." [Fearn, 2005]

As a specific example, an array of nine 40 cm beam diameter thrusters using xenon propellant and operating at 10 kV will consume 7.4 MW. The total beam current will be 702 amperes, for a thrust of 120 Newtons, and an Isp of 11,100 s.

This cluster of four ion thrusters shows that individual engines can combine to produce higher total thrusts. However, if the Isps of the engines remain fixed, the total engine cluster will consume fuel four times faster than an individual ion engine. But if you increase the Isp of each engine by a factor of four, you only need the same total fuel of one thruster. (Hall Effect thrusters. Credit: Busek/Space.com)

The Japanese Hayabusha (Muses-C) spacecraft used four ion thrusters, with three in operation at any one time. The dry mass of the Hayabusha engines is 59 kg including a gimbal and a propellant tank, which was filled with xenon propellant of 66 kg. A single engine is rated at 0.008 Newtons thrust, 3,000 sec Isp, and 350 W electrical power consumption, so that the Hayabusa spacecraft is accelerated 4 m/s per day by the maximum thrust 0.024 Newtons. It rendezvoused with the asteroid Itokawa after the deep space flight of two years with a delta-V of 1,400 m/s, 22 kg of xenon propellant consumption and 25,800 hour total accumulated operational time of the ion engines. The engine cluster generated a maximal thrust of 0.025 Newtons while consuming 1.1 kW of electrical power.

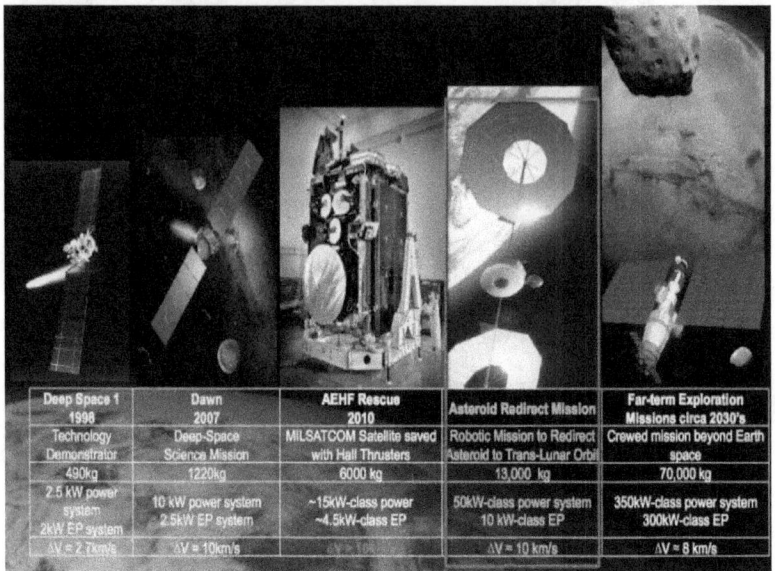

Deep Space 1 1998	Dawn 2007	AEHF Rescue 2010	Asteroid Redirect Mission	Far-term Exploration Missions circa 2030's
Technology Demonstrator	Deep-Space Science Mission	MILSATCOM Satellite saved with Hall Thrusters	Robotic Mission to Redirect Asteroid to Trans-Lunar Orbit	Crewed mission beyond Earth space
490kg	1220kg	6000 kg	13,000 kg	70,000 kg
2.5 kW power system 2kW EP system	10 kW power system 2.5kW EP system	~15kW-class power ~4.5kW-class EP	50kW-class power system 10 kW-class EP	350kW-class power system 300kW-class EP
ΔV ≈ 2.7km/s	ΔV ≈ 10km/s		ΔV ≈ 10 km/s	ΔV ≈ 8 km/s

The Journey to Mars

Amidst all of the possibilities for getting to Mars in the next few decades, at the time of this writing, NASA has now developed a plan for getting to Mars that will involve ion engines powered by solar electricity rather than the more costly and politically riskier nuclear thermal engine technology. This is an incremental approach that is totally in keeping with NASA's careful and measured approach to using new technology ion space. Ion propulsion has been space-tested with the Deep Space I and Dawn missions, and on numerous Earth-orbiting satellites for station-keeping.

The performance of an ion engine is directly related to the scale of the electrical power system you can deploy, and currently solar-electric systems are the dominant power plants in the inner solar system out to the orbit of Jupiter (Juno Mission). The table above (Credit:NASA) gives one sense of how this planning works, and the scale of the ion engine performance that is needed at each step.

Some of the advanced photo-electric system being developed by NASA and NASA contractors are based on the solar energy technology used in the NASA Deep Space 1 mission and the Naval Research Laboratory's TacSat 4 reconnaissance satellite, and are based on 'stretched lens array' lens concentrators for sunlight that amplify the sunlight by up to 8 times (called eight-sun systems). The solar arrays are also flexible and can be rolled out like a curtain. The technology promises to reach efficiency levels of 1000 watts/kg, and less than $50/watt, compared to the 100 w/kg and $400/watt of current 'one sun' systems that do not use lens concentrators. A 350 kW solar-electric engine system is a suggested system for a 70 ton crewed mission to Mars. With the most efficient stretched lens array solar arrays currently under design, a 350 kW system would have a mass of only 350 kg and cost about $18 million.

As to what kind of ion engine technology will be used remains an open question. A VASMIR ion engine is still to be tested on the International Space Station, but the Mars plan being proposed requires Hall-Effect engines. In 2022, a Mars Orbiter mission to replace the aging communications network at Mars will probably be the first to deploy a solar-electric ion engine to test out the new high-efficiency solar panel and engine designs that will play a huge role in future manned trips to asteroids and Mars.

"An orbiter equipped with ion engines — and highly-efficient solar panels to generate the power for the thrusters — would likely launch into a high-altitude Earth orbit, then fire its electric rocket jets to gradually break free of Earth's gravity and fly to Mars. Instead of the usual transit time of less than a year, it might take two years for a probe with ion propulsion to get to the red planet, where it would initially enter a looping orbit far from Mars, before using its electric engines spiral closer." (Spaceflight Now)

Interplanetary Travel

These kinds of transit times are OK for unmanned systems but potentially lethal for manned systems. To dramatically reduce transit times using a solar-electric system, NASA is looking at the VASIMR engine technology developed by *AdAstra,* which we previously discussed. With promises of 39-day transit times, it seems like a conservative match to mission needs without 'going nuclear', however with present solar panel technology it is not at all clear, as its detractors have noted, how to generate the required power while keeping the mass of the solar electric system small compared to payload mass.

In the coming years, we will be hearing a lot more about NASA's plans for solar-electric ion propulsion as the non-nuclear alternative to interplanetary travel. It seems that the heavier but more efficient nuclear engine systems will have to wait until later in the 21st century before they will come into their own, politically.

Fusion-based engines

Although we are literally a stones-throw away from having honest-to-god fission-powered ion rockets plying interplanetary space just like most science fiction stories demand, we are nowhere nears stepping up to fusion drive technology.

The main problem is, 'What do we mean by fusion drive?'. Our encounter with fission-ion technology suggests that fusion is just a better way to generate lots of power. The more power we can generate with the fewer number of kilograms of reactor mass, the higher we can push the grid voltages, the propellant ejection

speed and our SI, and that translates directly into more payload mass delivered to our destination, and at a faster pace.

Interplanetary Travel

Currently there are several possibilities for generating electrical power in space. Solar panels deliver about 60 watts/kg in the inner solar system, and fewer than 3 watts/kg beyond Jupiter. RTGs deliver about 2 to 5 watts/kg and are heavily used in outer solar system missions like the Voyagers and Cassini. Then we have some fission rocket designs that deliver about 30 watts/kg. It is hoped that the specific energy of fusion-based designs will be around 3000 watts/kg, which will enable 30-day missions to Mars using SI=5000 ion engines!

NASA's Glenn Research Center has proposed a spherical torus fusion reactor for its "Discovery II" conceptual vehicle design. "Discovery II" could deliver a manned 172 000-kilogram payload to Jupiter in 118 days using 860 tons of hydrogen propellant, plus 11 tons of Helium-3-Deuterium (D-He3) fusion fuel. The hydrogen is heated by the fusion plasma debris to increase thrust with an exhaust velocity up to 463 km/s.

In the 1980s, Lawrence Livermore National Laboratory and NASA studied the "Vehicle for Interplanetary Transport Applications" (VISTA). The conical VISTA spacecraft could deliver a 100-ton payload to Mars orbit and return to Earth in 130 days. It used 41 tons of deuterium/tritium (D-T) fusion fuel and 4,124 tons of hydrogen propellant. The exhaust velocity would be 157 km/s.

The NASA/MSFC Human Outer Planets Exploration (HOPE) group has investigated a manned MTF propulsion spacecraft capable of delivering a 164-ton payload to Jupiter's moon Callisto using 165 tons of propellant (hydrogen plus either D-T or D-He3 fusion fuel) in about 330 days.

The University of Illinois has defined a 500-ton "Fusion Ship II" concept capable of delivering a 100-ton manned payload to Jupiter's moon Europa in 210 days. Fusion Ship II utilizes ion rocket thrusters (343 km/s exhaust velocity and Isp= 35,000 sec) powered by ten D-He3 IEC fusion reactors. The concept would need 300 tons of argon propellant for a 1-year round trip to the Jupiter system.

Princeton Satellite Systems and the Princeton Plasma Physics Laboratory are collaborating on a Direct Fusion Drive (DFD), which will be a D–3He fueled fusion powered rocket engine. The design is based on the Princeton Field-Reversed Configuration Reactor (PFRC) concept for a compact, clean, steady-state fusion reactor for power levels up to10 MW producing a thrust of 30 N at a specific impulse of 20,000 s.

A variety of fusion propulsion concepts were surveyed by C. H. Williams and S. Borowski (NASA Glenn), 1997: The basis of the survey was the NASA SEI (Space Exploration Initiative) report (1988-1992) The general consensus was that nuclear propulsion technology could shorten trip times and increase payload mass fractions by several orders-of-magnitude, which would be the key to outer solar system explorations.

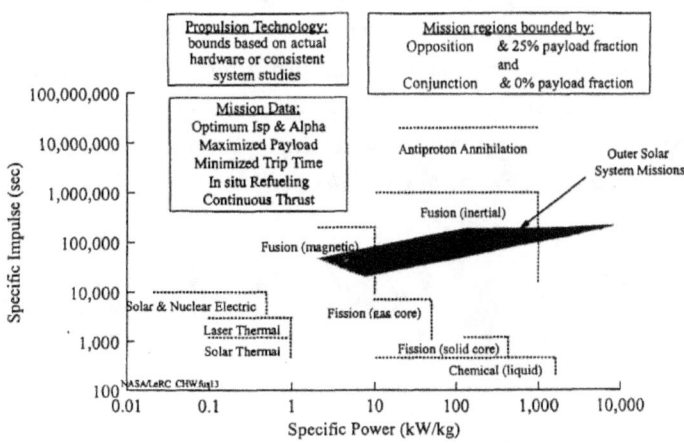

Engineers at the University of Washington are working on a fusion-powered rocket that could slash the estimated four-year round trip from Earth to Mars to a maximum of 90 days. The project has received two rounds of funding from NASA's Innovative Advanced Concepts Program totaling about $600,000. In their design, about 80 tons of lithium would be needed, with a payload mass delivered to Mars of about 80 tons. All portions of the design have been tested successfully in the lab, and the team now has to combine all the tests into a final experiment that actually produces fusion using its technology.

Z-pinch based nuclear fusion propulsion system may also lead to a small, low cost fusion reactor/engine assembly. The magnetic field resulting from the large current compresses the plasma to fusion conditions, and this process can be pulsed over short microsecond timescales. This type of plasma formation is widely used in the field of Nuclear Weapon Effects testing in the defense industry, as well as in the fusion energy research. An Isp of 19,436 s and a thrust of 3812 N per pulse has been predicted for this system.

Nuclear fusion rockets could slash travel times through deep space dramatically, potentially opening up vast swathes of the solar system to human exploration, said John Grunsfeld, associate administrator for NASA's Science Mission Directorate. *"It's transformative,"* Grunsfeld said last month after his presentation at Maker Faire Bay Area in San Mateo, Calif., a two-day celebration of DIY science, technology and engineering. *"You could get to Saturn in a couple of months. How fantastic would that be?"*

Nuclear fusion rocket design (Image credit: Prof. John Slough. University of Washington)

Other Propulsion Ideas

Solar Sail – Solar sails (also called light sails or photon sails) are a form of spacecraft propulsion using the radiation pressure (also called solar pressure) from stars to push large ultra-thin mirrors to high speeds.

Solar sails use a phenomenon that has a proven, measured effect on spacecraft. Solar pressure affects all spacecraft, whether in interplanetary space or in orbit around a planet or small body. A typical spacecraft going to Mars, for example, will be displaced thousands of kilometers by solar pressure, so the effects must be accounted for in trajectory planning, which has been done since the time of the earliest interplanetary spacecraft of the

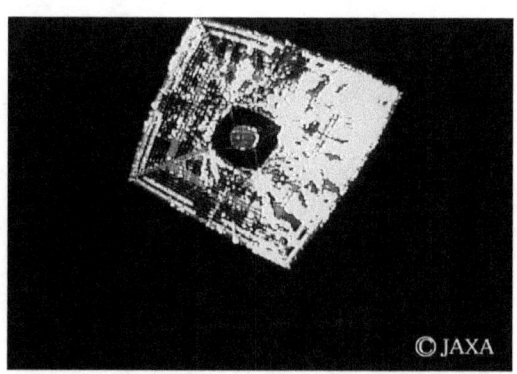

© JAXA

1960s. IKAROS, launched in 2010, has been the first practical solar sail vehicle. In 2012, it was still under thrust, proving the practicality of a solar sail for long-duration missions.

IKAROS has a square sail 14×14 m (196 m2) made of a 7.5-micrometre (0.0075 mm) thick sheet of polyimide, whichhad a mass of about 10 grams per square meter. IKAROS spent six months traveling to Venus, and then began a three-year journey to the far side of the Sun.

One of the earliest American stories about light sails is "*The Lady Who Sailed the Soul*" by Cordwainer Smith, which was published in 1960. In it, a tragedy results from the slowness of interstellar travel by this method. Another example is the 1962 story "*Gateway to Strangeness*" (also known as "Sail 25") by Jack Vance, in which the outward direction of propulsion poses a life-threatening dilemma. Also in early 20th century literature, Pierre Boulle's *Planet of the Apes* novel starts with a couple floating in space on a ship propelled and maneuvered by light sails. In Larry Niven and Jerry Pournelle's *The Mote in God's Eye*, a sail is used as a brake and a weapon.

Both Arthur C. Clarke and Poul Anderson, independently, but simultaneously, published distinct stories titled "*Sunjammer*" in March 1964. Clarke's depicted a "yacht race" between solar sail spacecraft, while Anderson, writing as Winston P. Sanders, depicts a maintenance crew, servicing space-freighters powered by light sails. Clarke published in the March 1964 issue of Boys' Life, while Anderson published in the April 1964 (on sale March 12 1964) issue of Analog Science Fiction /

Interplanetary Travel

Science Fact. In the episode "*Explorers*" of Star Trek: Deep Space Nine that aired in 1995, a reconstructed, "ancient" Bajoran "light ship" was featured. It was designed to use solar wind to fly out of a Solar System with no engine. In the film *Star Wars Episode II: Attack of the Clones* one is used by Count Dooku to propel himself across space. A solar sail was also used in James Cameron's *Avatar*.

The Planetary Society's LightSail satellite is a technology demonstration for using solar propulsion on CubeSats, which are a class of research spacecraft called nanosatellites. This first LightSail mission specifically is designed to test the spacecraft's critical systems, including the deployment sequence for the Mylar solar sail, which measures 32 square meters (344 square feet). The Planetary Society is planning a second, full solar sailing demonstration flight for 2016.

LightSail™ is a citizen-funded project whose first mission in 2015 hopefully will pave the way for a second, full-fledged solar sailing demonstration in 2016. (Image credit: The Planetary Society)

Laser

Laser propulsion is a form of propulsion where the energy source is a ground-based laser system. The basic concepts underlying a photon-propelled "sail" propulsion system were developed by Eugene Sanger and the Hungarian physicist Georgii Marx. Propulsion concepts using laser-energized rockets were developed by Arthur Kantrowitz and Wolfgang Moekel in the 1970s. Laser propulsion systems may transfer momentum to a spacecraft in two different ways. The first way uses photon radiation pressure to drive momentum transfer and is the principle behind solar sails and laser sails. The second method uses the laser to help expel mass from the spacecraft as in a conventional rocket. This is the more frequently proposed method, but is fundamentally limited in final spacecraft velocities by the rocket equation.

Solar Wind

A magnetic sail or magsail is a proposed method of spacecraft propulsion that would use a static magnetic field to deflect charged particles radiated by the Sun as a plasma wind, and thus impart momentum to accelerate the spacecraft. In typical magnetic sail designs, the magnetic field is generated by a loop of superconducting wire. The solar wind is a continuous stream of plasma that flows outwards from the Sun: near the Earth's orbit, it contains several million protons and electrons per cubic meter and flows at 400 to 600 km/s (250 to 370 mi/s). The magnetic sail introduces a magnetic field into this plasma flow, which can deflect the particles from their original trajectory. The momentum of the particles is then transferred to the sail, leading to a thrust on the sail.

For a sail in the solar wind one AU away from the Sun, the field strength required to resist the dynamic pressure of the solar wind is 50 nT. Robert Zubrin's proposed magnetic sail design would create a bubble of space of 100 km in diameter (62 mi) where solar-wind ions are substantially deflected using a hoop 50 km (31 mi) in radius. The minimum mass of such a coil is constrained by material strength limitations at roughly 40 tons and it would generate 70 N (16 lbf) of thrust, giving a mass/thrust ratio of 600 kg/N.

A Vorlon ship with unknown but very cool technology! (Image credit: Babylon 5)

The Last Hundred Miles

OK. So you have finally reached your destination in a few days or weeks. Now what do you do?

The rocket technology that got you to your destination is probably not suited to landing on the destination object, except in the cases where you can literally 'dock' with the object at speeds of a few meters per second. For anything else, such as an object bigger than about 500 km, you need a different spacecraft design.

Many science fiction stories have solved this problem by using the Mother Ship-Shuttle approach. The Mother Ship is a massive interplanetary transport vehicle designed to ship a lot of mass to a destination in the shortest possible time. This ship contains one or more smaller vehicles that are specifically designed to land and take off from the destination object. They are often aerodynamically contoured, have either vertical take-off capability, or can operate as high-speed planes and land on runways.

Interplanetary Travel

In the popular movie *Aliens*, the massive 78,000-ton interstellar troop carrier called the Sulaco included eight 35-ton 'dropship' vehicles for making planet fall. In the TV series *Star Trek*, we have the mother ship Enterprise, also designed for interstellar travel, carried several 10-ton shuttle craft in a hangar area, which are designed to enter planetary atmospheres and land on the surface.

In the Real World, we also have experience with shuttle systems of one kind or another. As countless astronauts can now tell us, making the trip to the surface of a planet is not that hard. It has now been accomplished hundreds of times using Apollo-like capsules, and even sophisticated Space Shuttles. But these systems take advantage of a planet's dense atmosphere in their descent to the ground, and with the exception of the Space Shuttle are not designed to return to space after the landing. Only Mars and Titan as destinations have an atmosphere, and terrestrial Apollo-like capsules would not work in the far-thinner atmosphere of Mars. To get to the surface of Mars and Titan with manned shuttle craft, or all of the currently favored airless destinations, you need a rocket system that provides powered descent, and of course ascent. This was accomplished by the Lunar Excursion Modules (LEMs) used in the Apollo moon landings, so we have some experience with this technology. It is worthwhile to note that because the destinations

342

have no atmospheres, we do not have to worry about atmospheric friction and making the shuttle vehicle aerodynamically sound. It just needs to have enough reaction mass and thrust to slow the shuttle down for a vertical landing and then to launch it back to the mother ship. We have also used this mother ship-shuttle approach with Cassini-Huygens, which visited Titan, with the 2015 Rosetta-Philae system, which landed on the comet 67P/Churyumov–Gerasimenko, and with a variety of missions to Mars that deposited landers and rovers on its surface. But only in the case of the Apollo mission did the lander return to the mother ship for a return voyage to Earth.

Titan and Mars have atmospheres and present us with some difficult issues. It is relatively easy to land on these objects using aerodynamic breaking and parachutes (Huygens) and in the case of the Curiosity rover, a combination of aerobreaking, parachutes and powered descent. But to get back to the mother ship, the Martian surface gravity is high enough that you need a system that reaches escape speeds of 5 km/sec, which is a significant speed comparable to our own Earth (10 km/sec). This will require a significant launch facility and a high-thrust system not unlike the conventional rockets launched from Earth's surface. The Mars shuttle craft will have to be a full-fledged rocket system that makes a powered decent, lands vertically, and then is ready for a vertical launch at the end of the visit because there are no runways...at least yet. The escape speed for Titan is only 2.7 km/sec so the shuttle craft requirements are significantly less than for Mars. (Image credit: The Mars Society).

Both Mars and Titan require considerable quantities of fuel to achieve the necessary delta-V of 5 km/sec (Mars) and 2.7 km/sec (Titan) for the shuttle to return to space. One possibility is for the shuttle to only bring to the surface enough fuel for a one-way trip, and then mine or extract the fuel needed for the return flight.

Titan's surface is rich in organic chemicals that can be 'cracked' to produce simple rocket fuels like liquid hydrogen and oxygen, however the process requires energy and only nuclear forms of energy generation could be used on the surface of Titan where solar electricity would be very difficult to generate in enough quantities. A nuclear system would be very heavy, so creating fuels in situ would be difficult. For Mars, the situation is much better with abundant solar energy and surface compounds known to be rich in trapped water, which can be cracked using simple electrolysis equipment. In fact, NASA is seriously looking into the option of creating a robotically-operated fuel depot on Mars that will run continuously and provide ready-to-go fuel for visiting explorers.

In 1997, the Reference Mission of the NASA Mars Exploration Study Team was produced for NASA (Stephen Hoffman, David Kaplan, NASAA Special Publication 6107), which specifically looks at a Mars Direct concept called In-Situ Resource Utilization (ISRU) where fuel is generated on the surface of Mars to supply the return flight to Earth. *"The hardware necessary to produce and store propellants using raw materials available on Mars (in this case, carbon dioxide from the atmosphere) is less massive than the propellant needed to depart the martian surface for orbit. It is now apparent that the technology for producing methane and liquid oxygen from the martian atmosphere and some nominal hydrogen feedstock from Earth is not only an effective performance enhancement but also appears to be technologically feasible within the next few years." (Section 3.5.3.5)* In 2021, NASA will launch a new Curiosity-style rover for Mars that will carry an instrument called MOXIE, which will make oxygen and test out the most important element of future technologies for extracting breathable gas and rocket fuel from the Mars environment. When operating, it will extract about ¾ ounce of oxygen each hour from the atmosphere. A scaled up version 100 times larger would be sent to Mars several years before any manned flight is launched from Earth, and stockpile enough oxygen and hydrogen for the return flight.

Many interesting destinations have no atmospheres at all, and present an even simpler design challenge for shuttle craft. Let's look at the only successful, space-tested design that has flown so  far: The Apollo LEM module. The mother ship was the Apollo Command Module 'capsule', which docked with the LEM at the end of the mission and brought the astronauts back to Earth. The LEM had its own launch platform that deployed once the LEM landed, and assured a stable geometry for the launch back into space. (Image credit: NASA/Apollo 17)

Interplanetary Travel

The pressurized 2-man crew cabin volume was about 160 cubic feet with a 100% oxygen atmosphere. The LEM had a mass of 10,300 kg of which 8,200 kg was fuel. The descent system used an engine with an Isp=311 seconds, and provided a total delta-V of 2,500 m/sec. The descent system included a 4-wheeled rover and the platform for the ascent stage. The ascent vehicle had a mass of 4,700 kg of which 54% or 2,550 kg was fuel. The engines provided a total delta-V of 2,200 m/sec.

Here is a list of total delta-Vs (2x escape speed) required to land and take off from a variety of interesting objects around the solar system. We have included objects that are also hazardous to humans, such as the larger moons of Jupiter, because robotic missions and tele-operation are viable options for exploring these bodies from more remote, and safer, locations.

Object	Diameter (km)	Mass (kg)	Delta-V (m/s)
Ganymede	5,262	1.5E+23	5500
Titan	5,150	1.4E+23	5400
Io	3,600	8.9E+22	5100
Callisto	4,820	1.1E+23	4900
Moon	3,475	7.3E+22	4700
Europa	3,121	4.8E+22	4000
Triton	2,705	2.1E+22	2900
Pluto	2,372	1.3E+22	2400

Titania	1,576	3.5E+21	1500
Rhea	1,527	2.3E+21	1300
Charon	1,208	1.6E+21	1200
Iapetus	1,470	1.8E+21	1100
Dione	1,123	1.1E+21	1000
Ceres	945	8.9E+20	1000
Tethys	1,062	6.2E+20	800
Vesta	525	2.6E+20	720
Pallas	544	2.1E+20	640
Enceladus	504	1.1E+20	480
Miranda	472	6.6E+19	400
Phobos	22	1.0E+16	22
Deimos	15	2.0E+15	12
Halley's Comet	10	2.2E+14	5
Comet Tempel 1	6	7.5E+13	4
1 km asteroid	1	5.2E+12	2
Comet 67P	4	1.0E+13	2
100-m asteroid	0.1	5.2E+9	0.2
10-m asteroid	0.01	5.2E+6	0.02

The Apollo LEM had adequate delta-V to land and return from our moon. That means that all objects below the moon in this table can be reached by using nothing more complicated than the 40-year-old Apollo technology! Even Io, Callisto and Ganymede would not be a significant stretch, requiring only some additional fuel to make the journey. The only object that presents a challenge would be Titan with its dense atmosphere, which we have previously considered.

At the present time, the space agency's NASA, ESA, RFSA and JAXA are working very hard to develop robotic sample-return missions. Thus far, we have landed on Titan (ESA Huygens), Mars (NASA Viking plus rovers) Venus (Soviet Veneras), asteroid Eros (NASA NEAR Shoemaker) and comet 67P/Churyumov-Gerasimenko (ESA Rosetta/Philae), but have only returned some dust samples from the tail of comet Wild 2 (NASA Stardust), and asteroid 25143 Itokawa(JAXA Hayabusa). No sample return missions have yet been attempted for high-gravity objects such as Mars or our own moon. Planned sample return missions include asteroid 1999 JU3 (JAXA Hayabusa 2 in 2020), asteroid 101955 Bennu (NASA OSIRIS-REX, 2017) and the moon (Change-5, 2017) and the Martian moon Phobos (RFSA Fobos-Grunt, 2024). Sample return missions to Mars have been proposed by NASA since the early-2000s but have not been funded for development, and will probably not occur until the latter half of the 2020s.

Thus far, there has not been a rendezvous and landing on an asteroid or comet that has gone perfectly. The most recent Rosetta/Philae landed on a comet nucleus at a speed of 38 cm/sec (0.02 miles per hour!) and bounced three times before coming to rest at 15:43 GMT on November 12, 2014 in a shaded region several km from its intended landing area.

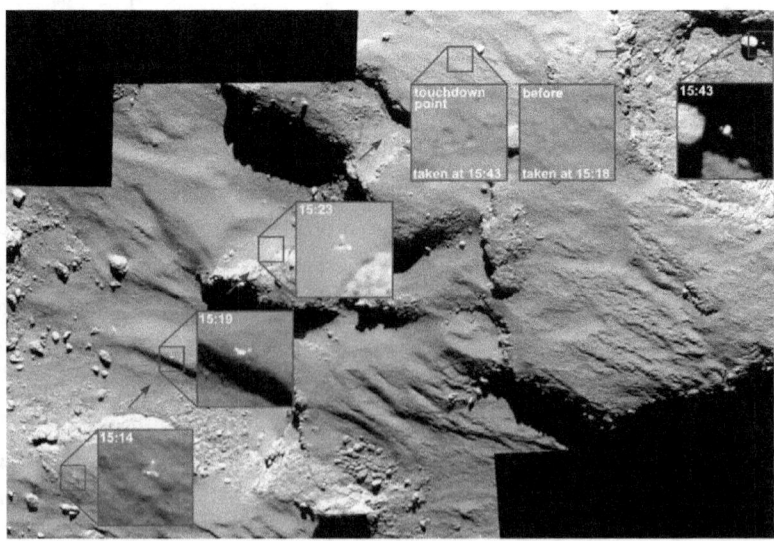

By the way, data from Philae suggest that, rather than being "soft and fluffy" as expected, the first touchdown site at 15:14 GMT contained a large amount of water ice under a layer of granular material about 10-inches deep. Philae detected the presence of molecules containing carbon and hydrogen it its atmosphere. Sixteen organic compounds were also detected, four of which were seen for the first time on a comet, including acetamide, acetone, methyl isocyanate and propionaldehyde. (Image credit:ESA/Rosetta)

Although landing on Mars and Titan will provide considerable challenges, already existing technology is well suited and tested for the ranges of delta-Vs needed for landing and returning from virtually all of the other interesting non-planetary locations in the solar system. The challenges we currently face involve designing and operating remote, autonomous landing vehicles on these surfaces, which will be eliminated as an issue when the vehicles are fully-manned and can anticipate more quickly the changing circumstances. (Image credit: NASA)

A Glimpse of the Near Future

Interplanetary travel will not be a stand-alone idea but will be carried out in the context of other issues that we confront as a society. What will those issues be in the next few centuries?

Today, as the beneficiaries of hundreds of years of scientific progress and mathematical computing ability, we can in fact make accurate predictions about some things such as when an eclipse will occur a thousand years from now, or what the sun is likely to be doing in a billion years. But for virtually anything that has to do with humans or human activity, we can barely anticipate accurately what is likely to happen in a week or a year from now.

The first thing we notice is that there are many 'End of the World' predictions that never came to pass between 2000 and 2015. The most spectacular of these was the December 21, 2012 end of the Mayan calendar event. Many books were written about this, the news media enjoyed covering it, and many children were scared about it. By some counts, since the 1800s there have been no fewer than 30 of these End of the World predictions and all, of course, have failed to materialize. Yet there are huge communities of adult humans who just love to believe in this concept and that, furthermore, you only need a strong belief or faith in the credibility of the forecaster to go ahead and plan your lives accordingly. This has led to several tragic incidents like the Comet Hale-Bopp event where people took their own lives out of desperation and fear. And still we choose to believe that 'certain people' have a unique skill in making such a dramatic prediction because of a paranoid mistrust of authority.

The second thing we notice is the number of science fiction stories set during this time period, which of course never came to fruition either. The major author of this period was Arthur C. Clarke, with his *2001: A Space Odyssey* universe, and the second journey to Jupiter in 2010. The former story was written/produced in the late 1960s, and the latter sequel was written in 1982. Although it can be argued whether either story really captured the reality of the post-Soviet world, it is quite clear that we never amassed the resources to build the massive Clavius Colony by 2001, or perfected even a rudimentary interplanetary travel capability with suspended animation and nuclear rockets. It is amazing to be writing a story in 1982 that so badly captured the reality of life in the early 21st century a few decades later, but then again, the genera is called science FICTION, and so it has to be faithful to the universe that it creates. I still watched both of these movies five times each! But science fiction is our only 'formal' guide to the future written by talented story tellers who honestly try to extrapolate from where we are to where we as humans might wind up.

The third thing we notice is that most solid predictions have to do with astronomical events. Planets are dumb, and their lives and behavior are controlled by only one thing – gravity. Once we figured out how gravity works back in the 1700s, we could launch the most amazing predictions if we were willing to literally crank out the calculations step by step. Once computers were available for this purpose in the 1950s, our forecasts became not only faster to generate for the immediate future (Nautical Ephemerides), but we could cast our predictive nets thousands of years backward and forward in time, and be confident that these predictions actually meant something. This trend continues into our distant future until we at long last run up against a major problem that defeats any further calculations of planetary configuration: poor-quality data and round-off error, but we will discuss these later.

In the following, I have used the solar system calculations by John Walker who used a sophisticated model of the solar system with time steps of 860 seconds and involved over 2 trillion calculations. This resulted in over 1 million transit events among the planets out to Neptune.

What do we make of predictions for 2015 and beyond? Again it depends on which domain you are curious about. We still have a very good handle on astronomical predictions, but now we get to include predictions that actually matter to humans. The most important of these is, when will the next large asteroid or comet impact Earth and possibly cause damage and kill people? What will climate change really be like?

As of the end of 2014, astronomers have systematically created a list of Near Earth Objects (NEOs) that are asteroids bigger than 100 meters, which have orbits that are close or intersect with Earth's orbit. The list is updated every month or so as new objects are discovered and their orbits analyzed for the next few hundred years. Each one is evaluated for the closest distance it will come to Earth in each of its passes, but as new data come in, these impact predictions get revised as the orbit is re-measured every year with new data. In 2004, the 300-meter asteroid 99942 Apophis was discovered, and what was known of its orbit at that time gave it a 3% chance of impacting Earth 2029. Follow-up measurements finally eliminated this asteroid as an impact risk in 2029, and reduced its impact risk on April 13, 2036 to 1 chance in 45,000.

The search for more of these NEOs continues on a shoestring budget around the world, with dozens of new objects detected every month and their orbits dutifully calculated for the next centuries. But as the Chelyabinsk Event in 2013 shows, some of these 50-meter objects can still sneak by if they come at us from the day-side and close to the sun, and they can certainly cause city-wide damage and injury.

For non-astronomical predictions there are many, but as we get father from today, their numbers greatly decrease because when it comes to human activities, who in their right minds really believes that you can predict human behavior or activity decades into the future? The near-mythological standing of the 21st century from the vantage point of the early 1970s shows how consistently bad we are at doing this for an activity as major as space travel, let alone what forms of energy we will be using to drive our civilization. We do not have flying cars and elaborate lunar colonies…yet.

I will list the science fiction story years in italic font followed by the initials of the authors, but will highlight in bold text those predictions that are the most secure; almost all being astronomical or mathematical certainties. Note, we can't even list NASA launch dates with certainty for missions that can still suffer from calendar slippage. An example of this is the Webb Space Telescope – one of the most expensive science missions that NASA has undertaken so far, and periodically subject to Congressional budget cuts.

2016 Launch of the Webb Space Telescope [2013:NASA]

2016 Transit of Mercury on May 9

2016 Internet traffic reaches 1 zettabyte per year. (Cisco:2014)

2016 Juno spacecraft arrives Jupiter, July 4 [NASA:2011]

2016 **Venus Jupiter conjunction.** Separation only 0.06 degrees August 26.

2016 End of summer sea ice in the Arctic (2012:The Guardian)

2017 Mobile phones used by 69% of world population. Half of these will be smartphones. 91% of internet traffic will be through mobile phones.

2017 Cadillac releases first driverless car (NBC News: 2014)

2017 Boeing certified to transport astronauts to space station (Boeing:2014)

2017 **Total Solar Eclipse in continental USA, August 21.**

2018 World Wide Web traffic reaches 400 terabytes per second. A million minutes of video every second. 1.6 zettabytes per year. (Cisco:2014)

2018 **Next Orion test flight by NASA.**

2018 **Double Blue Moon in 1 year.** January 31 and March 31

2018 Launch of Webb Space Telescope in October [2014:NASA]

2018 Hayabusa 2 arrives at asteroid 1999 JU3.

2018 First un-manned mission to Mars launched (MarsOne.com: 2014)

2018 Launch of ESA ExoMars rover mission to Mars. (ESA:2014)

2018 China launches space station (China News:2015)

2018 NASA SLS deep-space rocket maiden flight. (PhysicOrg:2014)

2019 *Blade Runner movie events begin (PKD)*

2019 *Cities in Flight. Discovery of gravitron 'Spindizzy'. (James Blish)*

2019 Next launch of the IXV-European Space Plane. (ESA:2014)

2019 Over 100 million SmartWatches in use. (Juniper Research:2014)

2019 NASA captures first asteroid. (NASA:2014)

2019 **Transit of Mercury on November 11.**

2019 Large Synoptic Survey Telescope begins operation. Generates 6 petabytes of data every year. Uses a 3 gigapixel camera to photograph the sky.

2020 **Saturn and Jupiter conjunction December 21 and 7 arcmin apart.**

2020 **USA phases out all general purpose incandescent bulbs.**

2020   Mars human landing (NASA, 2001) Sometime in 2030s [2010: NASA]

2020 Half of US children are members of a minority group. (NPR:2015)

2020   Apple, Inc debuts first electric car. (Bloomberg:2015)

2020   GOOGLE, Inc debuts first electric car. (GOOGLE:2015)

2020   The amount of information stored every year reaches 35 zettabytes. (IDC Digital Universe:2014)

2020 NASA Mars car-sized rover mission launch. (NASA:2015)

2020   Massive and uncontrolled deforestation in Africa (2000:African Development Bank) Deforestation slowing significantly [2013: BBC News]

2020   World population 8 billion (1995:John Tanton). Expected to top 8 billion by 2025 (2013:Voice of America)

2020   Electricity consumption in developing world=developed world (2002:Physics Today). This occurred in 2010 [2013: US EIA]

2020   *Podkayne of Mars (RH)*

2020   Kilamanjaro loses snow (Science, October 16, 2002). About 85% of glaciers were gone by 2007 [2009: NAS]. Probably not until after 2060. [2012: NASA Goddard]

2020 Peak of World Coal production. [2014:Royal Dutch Shell]

2020 US Census forecasts 334 million in USA [2014]. Currently in 2015 it is 321 million

2020 **December 21, closest Jupiter - Saturn occultation 6-arcmin separation.**

2021 **Blue Moon August 22.**

2021 Bangladesh goes 100% solar (Inhabitat.com:2015)

2021 Giant Magellan Telescope begins operation in Chile. Mirror diameter 22 meters.

2021 United Arab Emirates plans unmanned mission to Mars (TechTimes:2014)

2021 Humans sent to visit asteroid (NASA:2013)

2021 First crewed Orion flight by NASA. Called Exploration Mission-2.

2022 *Soylent Green events begin.* (HH)

2022 NASA launches ion-powered Mars Orbiter mission. (NASA:2015)

2022 Thirty Meter Telescope in Hawaii begins operation.

2022 HTML-5 debuts. (Webmonkey.com:2008)

2022 Space-X 'Dragon' module equipped for Mars sample return mission.

2023 **Blue Moon August 21**

2023 Elon Musk offers driverless cars with D-series Tesla sedan. (Huffington Post:2014)

2023 First artificial brain developed by the European Human Brain Project. (EETimes:2014)

2023 US Army robots outnumber human soldiers by 10 to 1. (Gizmodo.com:2013)

2023 US debuts first exaflop supercomputer. One exaflop equals one quintillion (a quintillion is 1 followed by 18 zeros) calculations per second. It is the next great goal in supercomputing that followed the U.S. achievement in 2008 of reaching one petaflop, or 1,000 teraflops, on a system built by IBM. A petaflop equals one quadrillion (1 followed by 15 zeros) calculations per second. The 2023 date "is when we are going to have an exascale system," William Harrod, Research Division Director for DOE's Advanced Scientific Computing Research program, said in an interview. While the U.S. has spent about $300 million so far on the next generation of systems, that's a "low level," said Harrod. (ComputerWorld:2014)

2023 US Federal debt grows to $26 trillion (CBO:2013)

2023 China now has the most billionaires in world. (rt.com:2014)

2023 First MarsOne 4-man human settlement on Mars. (MarsOne:2012)

2023 Julian Day number 2460000 arrives February 12

2024 European Extremely Large Telescope begins operation in Chile. It has a 39-meter equivalent segmented mirror system.

2024 Russia removes its ISS modules to form its own space station. (SpaceFlightNow:2014)

2024 The Square Kilometer Array generates 1 exabyte of astronomical data every day.

2024 Hubble Space Telescope reenters atmosphere (2013)

2024 Total solar eclipse in USA April 8.

2024 Blue Moon August 19.

2024 ISS funding by US extended by President Obama to at least 2024. (NASA:2014)

2024 Orion Asteroid Mission unlikely to occur before now. (NASASpaceflight:2014)

2024 Holographic ticket agents at the airport. (Skyscan:2015)

2024 Privately-funded moon mission to drill into crust. The Lunar Mission One will drill 65 feet below the South Polar crust.(Space.com:2014)

2024 Earliest ISS retirement year (2013:Space.com).

2025 No dark skies anywhere in eastern US (2010: Dark Sky Association)

2025 World reaches 180 zettabytes of information produced each year. Mobile data traffic at 20 exabytes/month. 50 billion connected devices including 8 billion smartphones. This equals 40 billion years of high-definition video. At $80 per terabyte today, to store this much information would cost $80 billion. (FierceWireless.com: 2014)

2025 One third of all jobs done by robots or software. (PC Magazine:2014)

2025 Elon Musk predicts millions of electric Tesla cars produced every year.

2025 **Peak of sunspot cycle 25 based on reliable 11-year cycle**

2025 *Childhood's End (Arthur C. Clarke).*

2026 **Blue Moon May 31**

2026 Elon Musk (Space-X) predicts human on Mars by 2026.

2026 **Asteroid Apophis will not collide with Earth**. (NASA:2013)

2027 **Blue Moon May 20**

2028   ISS reenters atmosphere (2013:NASAspaceflight.com)

2028 Blue Moon December 31

2028 Astronauts land on lunar far side. (NASA:2013)

2028 Launch of Athena X-Ray telescope by ESA. (Space,com: 2013)

2028   **One-mile asteroid 1997 XF11 passes within 600,000 miles of Earth. (Spacewatch:1998)**

2029   Peak of world gas production [2014:Political Economist]

2029   **Apophis closest approach April 13, 19,000 miles to Earth's surface.**

2029 Blue Moon, August 24.

2029 Harvest Moon falls on autumnal equinox.

2029   The Singularity arrives where computers are as complex as human brains. From there, it is proposed, machines will be able to outcompete humans at just about everything. There will be superior car-parking algorithms, disaster-relief coordination, legal briefs, rocket science, great thinking in general. Taking it to the next logical step, this artificial intelligence will have to be managed carefully, because after all, any of these future brain machines will outthink and outmaneuver us at every practical turn. [2014: Bill Nye]

2029 Moving a family to Mars could cost $500,000 (Elon Musk:2012)

2030 Forty world cities with populations of more than 10 million.

2030 Ninety percent of news stories written by computers. (Infowars.com:2014)

2030 Russians begin colonization of the moon (Russia Today:2014)

2030 Retirement Apocalypse. Baby Boomers exhaust Social Security while Generation-X begins to retire (Time:2014)

2030 Insolvancy of US Medicare Trust Fund (2002; Washington Post). Will last until at least 2026 [2013: US Treasury Department]

2030 *The Star Dwellers. Haertel drive invented (James Blish)*

2031 US could put humans on mars by now. (Aldrin:2009)

2032 NEO 2013 TV135 410-meter. Collision likely.

2032 **Transit of Mercury, November 13.**

2033 *The Light of Other Days. The Wormwood asteroid is detected.*

2033 **Unix Date 2000000000 on May 5.**

2033 Humans arrive at Mars (NASA:2012)

2033 Social Security Trust Fund runs dry (PBS:2012)

2034 Half of all jobs will be automated (Huffington Post:2014)

2034 Over 1 billion people carried by US airlines. (FAA:2014)

2034 Four-year college degree could cost $422,230. (TheDaily: 2014)

2034 Swiss shut down all of their nuclear reactors. (Minister of Energy:2011)

2034 We will discover extraterrestrial life by now (SETI: 2014)

2034 Primary source of world electricity will be natural gas. (Scientific American:2010)

2035 Oil production and discover will peak before this year. Surveys of oil experts indicated that oil production will peak between 2010 and 2030, but certainly no later than 2040. Peak oil for Saudi Arabia is predicted for ca 2027. [2014:Wikipedia]

2035 Natural gas surpasses coal as source of electrical energy production [2014:US Energy Information Agency]

2035 Humans on Mars (NASA:2014)

2035 75% of new vehicles will be self-driving (Navigant Research:2014)

2035 Energy production surpasses consumption (BP:2014)

2035 Space elevator built (Extremetech.com:2014)

2035 No poor nations. Everyone at least lower-middle class (Bill Gates:2014)

2035 Arctic free of land-based pack ice in September (2014:FutureTimeline.net)

2035 Himalayan glaciers vanish (IPCC:2010)

2036 Peak of sunspot cycle 26 – if there is one! According to the 90-year Gleissberg cycle, the next solar maximum – ca 2024 – will probably be a dud too, but then cycles will become more energetic once again, and any cooling effect the brief downturn has had on Earth's climate will also vanish.

2036 *The Light of Other Days. Wormhole manipulation (Arthur C. Clarke)*

2036 November 8 is Excel serial Day 50000 since 1909.

2036 Earth crosses 'climate threshold'.(Scientific American:2014)

2036 Apophis 99942 will not collide with Earth. (NASA:2013)

2036 Asteroid Apophis may hit Earth- 390 meters wide (NASA:2008)

2037 NASA to put man on Mars (NASA:2007)

2037 US Debt will be twice its GDP (CBO:2012)

2038 **Unix 'Y2K' date crisis January 19**

2039 **Transit of Mercury, November 7.**

2039 US Debt exceeds entire US output GDP. It is about 106% of the GDP (CBO:2014)

2039 The first Trillionare.(Fiscal Times:2014)

2039 US Population reaches 400 million (US Census Bureau:2008)

2040 **This is the year 5800 in the Hebrew calendar.**

2040 Census predicts 380 million people in the US [2014]

2040 Latest Hubble reentry year (2013:Spaceflight Now)

2040 Bill Gates becomes the world's first Trillionaire. (Forbes:2015)

2040 Enough planetary systems have been searched that extraterrestrial signals should be detected. (SETI:2014)

2040 World faces water crisis.

2040 Arctic summers ice free (National Geograpic:2006)

2040 Sea level rise by 2.5 meters (Arctic News:2014)

Interplanetary Travel

2041   US Social Security Trust Fund runs out of cash (2002: Washington Post]

2042   *Timemaster. Expedition to Alpha Centauri* (Robert Forward)

2042   Known uranium reserves exhausted (2000: AltEnergy.org)

2042 Whites become the US minority (US Census:2014)

2043 Tourist trips to the Moon (Virgin Galactic:2013)

2043   *12 Monkies* year after the world plague of 2015.

2044   *Timemaster. Wormhole travel (*Robert Forward*)*

2044   *The Light of Other Days. WormCam invented (Arthur C. Clarke)*

2044 Driverless cars become mandatory (IEEE Spectrum:2014)

2045   *Rocheworld. (*Robert Forward*)*

2045 World population reaches 9 billion (UN:2011)

2045   If you bought a home in 2015 your 30-mortage is now paid up.

2045   Artificial Intelligence outmatches combined intelligence of humanity. (Business Insider:2014)

2045   Humans can upload their minds into a computer (Livescience:2014)

2047 Coldest year will be warmer than hottest year since 1860. Scientists from the University of Hawaii at Manoa calculated that by 2047, plus or minus five years, the average temperatures in each year will be hotter across most parts of the planet than they had been at those locations in any year between 1860 and 2005. (New York Times:2014)

2047 Constant heat waves make New York City a summer ghost town.(Bloomberg:2014)

2049 Transit of Mercury on May 7.

2049 Full moon occurs on Friday 13th, June. Not since June 2014.

2050 Americans over age 65 now totals 88 million [2010: US Census]

2050 World population now at 9.0 billion [UN:2014]

2050 *The Star Dwellers.* (James Blish)

2050 *Childhood's End.* (Arthur C. Clarke)

2050 Photovoltaics account for 18% world power (1997: PV Forum). Could generate 16% by 2050 [2014: International Energy Agency]

2050 GM sells 1 million fuel cell cars (1999: Road and Track). Carbon Trust predicts 690 million cars [2012: Carbon Trust]

2050. Computer speeds of 5 million GHz, if Moore's Law still works. [2010]

2050 World power 900 exaJoules per year (Science 2000)

2050 Humans will regularly reach 100+ lifespans and our reproductive years will be later and older and we will have fewer babies. First-time mothers in America were 21 in 1970 and 28 in 2011. We will have more intellectual play time and less time spent in robot and computer-controlled factories. Millennials spend more time in pre-adulthood than their grandparents and are far more technologically adept. [2014:Medical Daily]

2051 **Transit of Venus as viewed from the minor planet Ceres.**

2051 US population reaches 400 million (US Census Bureau:2014)

2052 **Transit of Mercury, November 9.**

2052 Nickle will finally be mined-out and in short supply. (Scribol.com:2014)

2053 *Third World War would have happened by now (Star Trek).*

2054 Arctic will be ice-free for several months each year (US News:2013)

2054 *The year of Minority Report movie.(2002)*

2054 Plant protean is now 30% of all protein in food (Lux Research:2014)

2055 First mass shooting occurs in a lunar colony (NASA:2013)

2055 There are now 1 billion Millionaires in the world. This equals 20% of the adult population. There will also be 11 Trillionaires. (Forbes and Credit Suisse bank:2015)

2057 Male-female wage gap finally eliminated at current rate of progress (ABCnews:2013)

2057 Life expectancy for women born this year is 100 years (DailyMail:2015)

2057 **The 100th anniversary of Sputnik satellite launch.**

2058 *Lost in Space movie.* Jupiter 2 is launched from Earth.

2060 Solar electricity produces most of world's power (Bloomberg:2011)

2060 Fully-driverless cars now widely available (Victoria Transporttion Policy Inst.:2015)

2060 Arctic sea ice disappears completely in summer. (NCAR: 2011)

2060 Dow Jones Stock index reaches 100,000 (TradingView:2014)

2060 The 'redhead' gene may be extinct due to genetic mixing (National Geographic:2014)

2061 *Star Trek. Warp drive invented by Zephram Cochran*

2061 *Babylon 5. First permanent lunar colony established.*

2061 July 28. Halley's Comet returns.

2061 Transit of Saturn from Neptune on May 29.

2062 *This is the year of the Jetsons TV Series.*

2064 Latest year for 50th US President to be elected if 8-year terms after 2016.

2065 *Procyon's Promise.Maker ship enters solar system (MM)*

2065 Ozone hole over Antarctica has fully recovered (ScienceNews:2005)

2065 Global warming begins to bankrupt world economy. (Rense.com: 2000)

2065 November 22 Venus transits Jupiter

2067 July 15 Venus transits Neptune

2068 Asteroid Apophis may collide with Earth (Christian Science Monitor:2009)

2069   This is the 100th Anniversary of Apollo 11 astronauts landing on the Moon.

2070 Complete disappearance of Northern Rockies snow pack (USGS:2003)

2070 Global population will peak at 9 billion (New Scientist:2001)

2072 *Hellstar. (MRSP)*

2075 *The Moon is a Harsh Mistress (Robert Heinlein RH)*

2075 Kindergartners in school in 2015 will be retiring.

2076 The Tricentennial of the United States.

2076 US population reaches 590 million (National Association of Realtors:2012)

2078 Transit of Mercury November 14.

2079 Year 2000 in Indian Civil Calendar

2079 Mercury occults Mars, August 11.

2080 Of 19 Winter Olympics venues, only 10 have reliable snow cover. (PBS/NOVA:2014)

2080 NY city has 9 F increase in temperature, 39 inches sea level rise, 6 heatwaves per year. (NYC Panel on Climate Change:2015)

Interplanetary Travel

2080 Over 3 billion people will not have access to clean drinking water. (IPCC:2008)

2081 Transit of Earth as viewed from minor planet Ceres.

2081 *Year of Harrison Bergeron movie by Kurt Vonegut.*

2082 *The Light of Other Days (ACC).*

2083 Events from *Alien* movie begins.

2084 Transit of Earth as seen from Mars (November 10)

2084 North America warms 8 F, with 800 ppm CO2 (NASA:2014)

2084 The movie *Total Recall.*

2087 *The Architects of Hyperspace. Alexander Zepos dies(TM)*

2088 Mercury occults Jupiter, October 27.

2090 US crude oil production ends (2012: U.Colorado)

2088 October 27 Mercury transits Jupiter

2089 Year that *Prometheus* takes place.

2090 Great Lakes water level 3 feet lower (2000: Globalchange.gov)

2093 *Starplex. Milky Way wormhole network (RS)*

2094 Mercury occults Jupiter, April 7.

2095 Transit of Mercury, May 8.

2098 Transit of Mercury, November 10.

2100 Two-thirds of world's 6,700 languages vanish. [2000: Vision.com]

2100 World population at 11 billion if current trends in 2014 continue. (UN: 2014)

2100 Carbon dioxide at 700 ppm; 2.5 F global temp increase since ca 1960. (1999: IPCC)

2100 *The Ring of Charon. Earth destroyed (RMA)*

2100 Sea level rises by nearly 2 feet. (2007; IPCC)

2100 One meter sea level rise and flooding on US east coast (2014: NOAA)

2101 *Babylon 5. First Martian base established.*

2102 Polaris reaches its closest position to Celestial Pole. March 24 at 0.4562 degrees from North Pole.

2109 Time capsule at Rutgers University opened, April 27.

2115   **One hundred years after 2015.**

2119   *Orphans of the Sky. Centauri Expedition (RH)*

2123   *Starwings. Earth Holocaust; star travel. (GP)*

2123   **Venus will transit across the face of Jupiter September 14.**

2125   *Methusela's Children (RH)*

2126   **Mercury occults Mars July 29.**

2132   **Julian Day 2500000 arrives August 18**

2134   **Halley's Comet returns March 27.**

2144   *Aliens movie events. [1980]*

2150   *Dr Who. Earth overrun by the alien 'robotic trashcans' called the Daleks. [1968]*

2150   **June 25 longest total solar eclipse since 1973; 7 min 14s**

2150.   World population of 11 billion (UN 1994)

2150   One of every 3 people 60 or older (UN 1999)

2115   *Childhood's End. Adult humans extinct (ACC)*

2117   **Transit of Venus (December 11)**

2125 **Transit of Venus (December 8)**

2125 *Childhood's End. The Earth destroyed (ACC)*

2130 *Rama Revealed. Alien artifact (ACC)*

2150 One of every 3 people will be aged 60 years or older. (1999: UN)

2156 *Babylon 5. Centauri civilization contacts earth.*

2163 **Transit of Earth seen from Mars on November 15**

2178 **Pluto returns to the place in its orbit where it was first discovered in 1930.**

2186 **July 16 longest total solar eclipse in 10,000 yrs. 7m 29s**

2188 **Transit of Jupiter as viewed from Neptune; August 8.**

2189 **Transit of Earth from Mars on May 10**

2190 *Seetee Ship. (JW)*

2197 *Rama II. Alien spacecraft enters solar system (ACC)*

2197 *The Engines of God. The Iapetus Expedition. (JM)*

2199 *The Matrix. World run by an artificial intelligence.*

Interplanetary Travel

We now come to the end of the Near Future. During this period, we were still trying to make predictions about the human future, energy production, population size and attempts at predicting how climate change will play itself out on global temperature and sea level. These are the only issues relevant to civilization that anyone has felt motivated to try to anticipate so far in the future.

Recognizing the futility of creating interesting science fiction stories that begin only a few decades out, writers have moved their stories to 2050 and beyond. Only a few Old School stories written in the mid to late-20th century still find the first decades of the 21st century interesting, and with the right technology to make entertaining reading.

Meanwhile, astronomers continue to have no problems forecasting eclipses and unusual planetary alignments with enormous accuracy to the day, hour and minute across the centuries. Their high frontier is now in predicting the orbits of Near Earth Objects that may collide with Earth, and getting enough high-quality data to make these predictions accurate enough to detect an Earth impact decades away. This is much like hitting a bulls-eye from a distance of a hundred kilometers!

The Far Future

The far future can be considered the time period when sensible and defensible predictions of what humans will be doing have come to an end, and now only astronomical, geological or science fiction predictions have something to say. Today, there is nothing to be gleaned from events that play themselves out many centuries from now that will help us with our immediate problems of survival.

Extending from the 23rd Century to the year 10,000 AD, this is a time period when science fiction stories still try to paint for us an interesting future. In particular the 3rd millennium is the universe of Star Trek and Babylon 5. Beyond this time, only a smattering of stories and themes have been considered. It is impossible to imagine what kinds of human motivations would drive people after 4000 AD in the 5th millennium. It is like trying to understand the mind of a Sumerian or an Egyptian. Technology has become such a 'civilizing influence' it is hard to understand future humans if we cannot even anticipate what kinds of technology make them who they are, and circumscribe the issues of the day. Many of these technologies may even be implanted, rendering us hybrid creatures of biology and technology.

Throughout this entire period, however, astronomical predictions continue to provide a fixed background against which human events play themselves out. We know today, all the eclipses and planetary configurations, but we do not know if humans will still find them interesting. By these distant centuries, we will know the orbits of every object large enough to cause damage to civilization when it impacts Earth. We are most of the way towards accomplishing this feat today. The question remains whether we will be in a technological position to do anything about a potential impactor before it arrives. This requires us to entertain the thought that, either we will have a robust space presence centuries from now, or we will not. In the former scenario, civilization is always saved from such impacts literally by the artifice of a *deus ex machina* as the Greeks called it, and continues on its historic course. In the latter case, we have lost this capability and now are victims of this vast cosmic roulette game unable to do anything to ward off these catastrophes. Sadly, thanks to the work of astronomers during the 21st century, we will have a detailed calendar of when actual 'end of the world' events will happen as some kind of archeological legacy delivered to future generations from out of their possibly incomprehensible, technological past.

Another issue is the survival of our current records and accumulated knowledge into these distant centuries. Already we are transcribing all our history into electronic files. The 21st century may become the new Dark Ages if none of this information survives more than a few centuries after it is recorded. We already know that vast quantities of data are being stored on DVD and CDroms have lifetimes only a decade or so before they are unreadable. Nine-track magnetic tapes that were commonplace data storage media in the 1970s are now so degraded that they cannot be read in most cases, moreover, the tape readers are either gathering dust in museums or have already been recycled for their scrap metal.

Theoretical physicist and futurist Michio Kaku predicts that in a mere 100 years, humanity will make the leap from a type zero civilization to a type I civilization on the Kardashev Scale. In other words, we'll become a species that can harness the entire sum of a planet's energy. Wielding such power, 26th-century humans will be masters of clean energy technologies such as fusion and solar power. Furthermore, they'll be able to manipulate planetary energy in order to control global climate. Physicist Freeman Dyson, on the other hand, estimates the leap to a type I civilization would occur within roughly 200 years in the 23rd century.

2200 Worst case 6 F temperature increase globally (1999)

2200 Carbon dioxide may reach 1000 ppm, and a climate tipping point occurs by now. [2010:NAS survey of climatologists]

2200 *Galaxies Like Grains of Sand. Procyon Colony settled. (BA)*

2209 *The Engines of God. Expedition to Quraqua (JM)*

2211 **Halley's Comet returns for 33rd time.**

2223 **December 2, Mars Jupiter occultation. 9-arcsec**

2233 *Captain Kirk born (Star Trek)*

2231 *The Engines of God. Omega Cloud. (JM)*

2218 *Star Trek. First contact with Klingon Empire.*

2240 **The millennial year 6000 in the Hebrew calendar.**

2245 *Babylon 5. The Earth-Minbari War*

2247 **Transit of Venus (June 11)**

2250 Coal reserves under US ground exhausted.

2254 *Babylon 5. The Babylon 4 station goes online and vanishes.*

2255 **Transit of Venus (June 9)**

2256 Centralia coalmine fire may be ending (2006)

2257 *Forbidden Planet movie*

2257 *Babylon 5. The Babylon 5 dedicated. Commander Sinclare*

2260 *Babylon 5. Sheridan engages the Shadows in first major battle.*

2261 *Babylon 5. The Shadow-Vorlon war at Coridian.*

2261 *Babylon 5. Mars and Earth are liberated*

2262 *Babylon 5. Commander Lochley in command of Babylon 5.*

2262 *Babylon 5. Centauri Prime falls to Narn.*

2263 *Babylon 5. 'River of Soals' movie.*

2263 *Fifth Element movie*

2264 *Star Trek. Captain Kirk begins the historic 'Five year Mission'*

2267 *Babylon 5. Drakh release plague on Earth.*

2268 **May 13 Transit of Earth from Mars**

2270 *Babylon 5. The Drahk Plague is cured.*

2274 *Logan's Run movie events.*

2281 *Babylon 5. Sheridan 'dies'. Babylon 5 is destroyed.*

2300 **Possible start of Ice Age.** If global warming heats the Earth high enough, the Greenland Ice Cap will melt, flooding the North Sea with fresh water that shuts off the Thermohaline Circulation. This cools the northern hemisphere enough to cause significant cooling in the Arctic Region and the start of a new cold era. The IPCC Third Assessment Report notes that *"even in models where the THC weakens, there is still a warming over Europe"*. Model runs in which the THC is forced to shut down so show cooling – locally up to 8 °C (14 °F)— although the largest anomalies occur over the North Atlantic, not over land. However, a thermohaline circulation shutdown could have other major consequences apart from cooling of Europe, such as an increase in major floods and storms, a collapse of plankton stocks, warming or rainfall changes in the tropics or Alaska and Antarctica (including those from intensified El Niño effect), more frequent and intense El Niño events, or an oceanic anoxic event — oxygen (O_2) below surface levels of the stagnant oceans becomes completely depleted — a probable cause of past mass extinction events.

2300 *The End of Eternity. Humans discover Temporal Field. (IA)*

2300 *Ringworld Engineers. Moa becomes a popular food. (LN)*

2300 Since 2000 the day is 6 milliseconds longer

2300 Global temperature could be 21 F warmer than 2000. Half the world uninhabitable. [2010:NAS]

2305 *Jean-Luc Picard born (Star Trek NG)*

2310 *Cities in Flight. The Earth-Vegan War (JB)*

2341 *The Man-Kazin War begins. (LN)*

2354 Transit of Jupiter as viewed from Neptune; October 6.

2360 Transit of Venus (December 13)

2362 Halley's Comet returns for 35th time.

2364 *Star Trek Next Generation. Captain Picard begins 'Mission'*

2367 *Star Trek Next Generation. The Borg attack the Federation.*

2368 Transit of Venus from Earth (December 10)

2368 Transit of Earth from Mars November 13

2369 *Star Trek Deep Space 9. Captain Sisko assumes command.*

2371 *Star Trek Voyager. Chronicles begin in Delta Quadrant.*

2375 *Cities in Flight. Earth cities launched into space (JB)*

2394 Transit of Earth from Mars May 10

2400 *Eon. Humans hollow out the asteroid Ceres (GB)*

2427 *Destiny's Road. Earth's interstellar mission (LN).*

2439 *The Sundered Worlds. Earth galactic empire(MM)*

2447 July 14 is Excel serial day 200000 since 1900.

2447 Transit of Earth from Mars November 17

2464 Latest year for 100th US President to be elected if 8-year terms after 2016.

2473 Transit of Earth from Mars May 13

2478 Mars Jupiter occultation August 29,

2487 *Buck Rogers TV series. Last of America's deep space probes.*

2490 Transit of Venus (June 12)

2492 **All 8 planets in same 90-degree quadrant of the sky**
2498. **Transit of Venus (June 10)**

What other technologies will shape the world of the 26th century? Futurist and author Adrian Berry believes the average human life span will reach 140 years and that the digital storage of human personalities will enable a kind of computerized immortality. Humans will farm the oceans, travel in starships and reside in both lunar and Martian colonies while robots explore the outer cosmos. (http://science.howstuffworks.com/environmental/earth/geology/earth-500-years.htm)

2512 *Antares Dawn (MM)*

2525 *Zager and Evans sing about: In the year 2525*

2534 *The Light of Other Days. Wormwood impact Earth (ACC)*

2540 *Brave New World (Aldous Huxley)*

2552 **Transit of Earth from Mars May 16**

2562 **This is the year 2000 in the Islamic calendar.**

2517 *Serenity* events

2583 *Chronicles of Riddik* events

2589 **Halley's Comet returns for 38th time.**

2600 *The End of Eternity. Time travel invented (IA)*

Epilog

Although interstellar travel presents challenges that are far beyond our current level of understanding and technology, interplanetary travel is entirely workable within our existing technological attainments. It is not a matter of technological know-how, but simply political will.

Compared to interstellar travel, which is a topic I covered at length in my previous book 'Interstellar Travel:An astronomer's view', interplanetary travel and colonization are within our technological horizon even today. Part of the reason is that we can focus on far fewer destinations. This is an unpleasant reality, but it is based upon some very hard facts-of-life.

The most important of these is that virtually all of the destinations beyond the orbit of Saturn are ice-bound worlds and moonlets. These provide no useful resources from which to build habitats, unless we go to the extreme expense of building them in the resource-rich inner solar system, and literally dragging them billions of miles to these remote destinations. Having done this, we still require supply ships to bring us replacement parts and food stuffs, although water and breathable atmospheres can be made from locally-abundant ices.

There will be trips to the outer solar system, but these will come about as a byproduct of fast interplanetary travel developed to support activities on Mars, the asteroids and other inner solar system destinations. Once you can get to Mars in a week (or less!), journeys to Uranus and beyond will be possible for hardier voyagers willing to spend months or more to get there. There will be huge commercial pressures among competing companies to shorten travel times by advancing new rocket technology at the lowest costs.

The inner solar system setting is much more exciting and human-friendly, with plenty of raw materials that can be mined from asteroids and the surfaces of Mars and the moon. The relationship between these mining activities and supplying Earth with them is problematical. Mining or fabrication technology has to be brought into space at a current launch cost of thousands of dollars per kilogram. Mining operations are also not inexpensive, so all of these costs have to be tacked on to the price-per-pound of the recovered ore. That means on Earth, mined extraterrestrial ores will be orders-of-magnitude more expensive compared to those extracted from the ground or from aggressive recycling. Some kind of resource extraction will certainly have to be accomplished by future colonists, but these resources will remain where they are and not brought back to Earth. That being the case, what is the economic value to Earth-bound companies if they can't even bring back what they mine and sell it to the teeming billions of consumers on Earth? This is a hard-nosed question that has to be answered, not by High Frontier Dreamers but by economic and business realists.

The issue of where we will go with interplanetary travel requires realistic thinking in the face of the likely economic value. Most of the likely destinations near Jupiter and Saturn are either 'radioactive' and deadly to humans (the four giant Galilean moons), or are poisonous and frigid (Titan). There is a huge curiosity value in seeing them first-hand that any child in grade school can tell you about, but it is pointless to invest in a simplistic tourism industry that would cater only to a vanishingly small fraction of the population who are also Explorers and in excellent health.

The other thing we have to deal with is the quality of life. Humans were adapted to live on the surface of a very colorful and biologically-rich planet. How will humans feel if, after traveling to a distant destination, they must permanently live cooped-up inside a sealed habitat, and only venture outside in spacesuits? Astronauts can do this today because they KNOW that in a few months they will be returning to their natural setting on the surface of Earth. But on Mars or the Moon, you have no place else to go. After a few weeks, you will have walked around your entire 'neck of the woods' at huge personal risk and that will be the end of your novel experience. There are no rain storms, changes of the seasons, or other natural events to enliven your daily or annual existence. This is a huge psychological burden, and it is not clear that unmedicated human personalities could withstand this kind of existence. If that is the case, our concept of colonizing Mars is all about a circulating population of humans who may spend a year on Mars and then return to Earth to 'recharge' their psychological batteries. Until we have colonies with as much variety as downtown New York City or Paris, boredom will be our biggest psychological hazard.

In terms of rocket technology, it is pretty clear the road that we will take in the next few decades. We will start with solar-electric ion engines that have 10 times the specific impulse of chemical rockets. They will be powered by large deployable photovoltaic films, which will supply the megawatts of power needed to drive spacecraft throughout the inner, sunlight-rich, solar system. Trips to Mars on the order of weeks or less are easily forseeable, and based on even current engine designs, there are no obvious technological roadblocks to developing such engines. Although small kilowatt ion engines have already been used on NASA spacecraft, these provide only an ounce of thrust per engine. VASMIR engines being readied for testing on the International Space Station in the next few years will deliver several pounds of thrust. In 2022, a spacecraft now being designed for Mars will test-out high efficiency solar arrays and the newer generation of ion engines operating at much higher thrusts than Dawn or Deep Space 1. It is also expected that the Asteroid Rendezvous Mission in the mod 2020s will test out an even more powerful ion engine. So, we are moving away from chemical rockets and taking the next step to full-scale ion engine technology for most future interplanetary missions through at least 2030. But what about nuclear?

We have seen how dramatically efficient nuclear rockets can be even above ion rockets. With specific impulses reaching 10,000 or more, these are truly the future workhorses of interplanetary travel and promise journeys to the outer planets taking days or weeks. But current designs are massive and dirty, and there is no political will among the Western Democracies for placing them on launch pads. Only China, with its weaker public control of risk-taking might be in a position to deploy nuclear rockets before the middle of this century. This places the realistic horizon for nuclear rockets well into the last half of the 21st century at best. For now, the much safer solar-electric systems will dominate. They have spin-off technologies that directly impact domestic green energy development.

Today, a home owner can cover the roof of their home with solar panels to generate the electricity they need. With the solar technology needed to run ion engines for quick interplanetary travel, such solar panels may only amount to a single square yard of roof surface! This is how interplanetary travel will be hastened through green energy development on the ground. This is unlike the development program for nuclear rockets, because there are no economic or commercial reasons to build small, low-mass portable nuclear reactors. That niche has been completely occupied by roof-top solar panels and a non-centralized strategy for energy production.

The opportunity for making interplanetary travel a reality is within our grasp in terms of technology. That is why I am so excited about the prospects for a Mars landing in the 2030s, and where that effort will take us in 2040 and beyond. Interplanetary travel will never be the family-style jaunt in space that author Robert Heinlein described in his many books, but it will steadily become more of a reality as time goes on. By the 22nd century, it is inconceivable that regularly-scheduled trips to Mars by explorers and scientists will not be commonplace. This may also be the century when some limited commercial tourist travel to the moon and Mars becomes possible as rocket technology advances and travel times become weeks and not years.

From that point on, there will be much that we can do in this solar system in the centuries that follow to erect actual cities on the moon and Mars near water-rich locations, and create human environments that are socially-complex, experientially-rich, and perhaps politically-independent.

Meanwhile, let's get started writing more science fiction stories that celebrate the many exciting possibilities within our own solar system.

The stars can wait for another millennium.

Image Credits

The History of Interplanetary Travel

-- X-20 Dynasoar on top of a rocket ca 1963. Aerojet.

--Refueling in space. IF magazine cover.

Where should we go?

-- Solar system diagram. NASA Space Place.com

--Image of Mercury with SOAR telescope - University of North Carolina,
G. Cecil from http://tinyurl.com/md2qmbe   Sean Walker,
October 8, 2007. gerald@thececils.org,

-- Mars telescope image, Pic-du-Midi Observatory

--Ganymede image. Damian Peach, -http://tinyurl.com/27wl9xk

--Near Earth Objects, Armagh Observatory http://tinyurl.com/2s6hvn

-- Asteroid orbit diagram. Smithsonian Minor Planets Center.

--Chelyabinsk Meteor. http://tinyurl.com/qzh9lmd

Why should we go?

--NASA funding since 1958. Wikipedia.

--NASA spinoff diagram. http://er.jsc.nasa.gov/seh/spinpost.gif

--The future - Gunter Radtke http://tinyurl.com/objhm3u

--Chinese Yutu rover. CNSA / CCTV

--NASAs ARM capturing an asteroid. Credit NASA.

--Design for Mars colony. Space-X.

--Spacesuit sewing. ILC (Playtex) Dover, LP) http://tinyurl.com/lv9k7t3

How much will it cost?

--NASA vs World bar graph - Statista.com http://tinyurl.com/nt97hx9

--US economy 2012 pie graph - http://tinyurl.com/q2uphag

--Image from The Matrix movie. http://tinyurl.com/nqrd3gz

--African village. http://tinyurl.com/ozojpxg

Some Common Ideas

--Components of GDP (Wikipedia) http://tinyurl.com/opx5dcf

--Growth and Debt chart (Reuters) http://tinyurl.com/72wwnnr

--Five things people buy - Mashable.com http://tinyurl.com/ouho6n2

--Bigelow Olympus habitat. http://tinyurl.com/nu8jyqj

Resources and Mining

--Mining on the moon -: Pat Rawlings/NASA/Zuma Press

--Moon's crust, maps - http://tinyurl.com/25xxwr

--Different asteroid and meteor types. Wikipedia.

--Asteroid mining. Artist concept by Denise Watt/ NASA

--NASA's proposed $2.6 billion asteroid capture mission. Image credit: Astrobob.

-- Surface of Comet 67P: ESA/Rosetta.

--Mercury Magnesium map NASA / JHU http://tinyurl.com/nsr3v8m

--Mercury Element ratio diagram. http://tinyurl.com/nv9qgab

--Mercury Surface composition map. http://tinyurl.com/k6getoy

Mars Resources

--Silicon map http://tinyurl.com/ptyy33x

--Thorium map http://tinyurl.com/nqvabpx

--Iron map http://tinyurl.com/pmap8em

 -- NASA/JPL/Max-Planck-Institute for Chemistry

-- NASA/JPL-Caltech/University of Guelph/CSA

-- Element abundances http://tinyurl.com/nb2lj8a

Asteroids, Comets and Moons

-- Ice on Europa map. http://tinyurl.com/nkrvxts

--Five views of Iapetus http://tinyurl.com/q3bvwym

--NASA/JPL/University of Colorado/Space Science Institute (PIA09970)

--UV image of Iapetus and ice deposit. NASA/JPL/U. Colorado

--Comet 67P http://sci.esa.int/rosetta/14615-comet-67p/

--http://www.hou.usra.edu/meetings/lpsc2015/pdf/2092.pdf

--Itokawa, courtesy of the Planetary Society.

--Itokawa close up. http://tinyurl.com/nsmtj9s

--Chemical composition of comets. http://tinyurl.com/njclpn3

--Europa map of water (NASA) http://tinyurl.com/pol85hq

--Iapetus composition strip. http://tinyurl.com/q3bvwym

--UV image of Iapetus. http://tinyurl.com/qxy9lt8

--Hydrogen map of Vesta surface http://tinyurl.com/pg4ddrc

--Image of comet http://sci.esa.int/rosetta/14615-comet-67p/

--Itokawa close up. http://www.psi.edu/pgwg/images/oct09image.html

When should we go?

--Seismic study of Earth's interior in 3-D. http://tinyurl.com/pf43dek

--Journey to Mars diagram (NASA) http://tinyurl.com/nqycn5k

-- National Geographic. http://tinyurl.com/oecrlk8

-- Europa Mission. NASA.

-- Space roadmap. NASA

--Mars habitats. NASA

Places to Visit

--Delta-V budget (Wikipedia) http://tinyurl.com/3hz5xpl

--Earth-Moon lagrange points, http://tinyurl.com/q4zpqs3

Interplanetary Travel

--Delta-Vs across solar system. http://tinyurl.com/qarecpc

-- Jupiter system. NASA/Science Visualization Service

-- Jupiter magnetosphere and moon

-- Jupiter moon orbits ssheppard@carnegiescience.edu -
http://tinyurl.com/obvryxx

-- Saturn moons Wikipedia/NASA

--Uranus moon orbits Scott Sheppard. http://tinyurl.com/psxyzd9

--Neptune moons Scott Sheppard. http://tinyurl.com/p2dld9b

--Saturn outer moons. Scott Sheppard http://tinyurl.com/nfvdhrx

--Pluto moons. Scott Sheppard. http://tinyurl.com/nfbhyxr

--Artist render from pluto. (Image credit: Wikipedia/ESO/L. Calcada)

--Amundsen-Scott   Station.   NSF/Jeremy   Johnson)
          http://tinyurl.com/o9rmrju

--Jupiter radiation belts http://tinyurl.com/3qa54ht

Orbital Platforms

--Phobos – NASA Mars Reconnaissance Orbiter.

--Deimos - NASA Mars Reconnaissance Orbiter.

--Pluto satellites – NASA/Hubble Space Telescope.

Interplanetary communication.

--Canberra DSN station NASA

--Voyager radio image -NASA/NRAO http://tinyurl.com/oq8ubj8

-- Laser ranging. Apache Point Observatory

--A ready-to-go laser communications transceiver. Image credit: JPL

-- Prototype of an interplanetary laser system. http://tinyurl.com/nsrqtaf

Radiation Hazards

145--Hand x-ray image: http://www.sunsetradiology.net/xrays.html)

149--Radiation dosage pie graph: http://tinyurl.com/mxg62xe

151--Cosmic ray spectrum. http://tinyurl.com/mghcdl9

153--Active shielding study. NASA: http://tinyurl.com/m6r9u6

--US radon map, http://tinyurl.com/kkrw8zh

Mutations

164--Population size and mutations. Cameron Smith.

165--Bar graph of radiation exposure: NASA/JPL-Caltech/SwRI

The Hazards of Interplanetary Space

130--The Student Dust Counter (SDC), NASA/New Horizons

131--Interplanetary dust particle: Donald E. Brownlee. Wikipedia

--Map of bolide events http://tinyurl.com/ntguba7

--Map of lunar impacts – http://tinyurl.com/6nb8g8

--Meteorite impact frequency –

-- Goldstone tracking dish DSN. http://tinyurl.com/npe2m53

141--Debris frequency figure: http://tinyurl.com/k8xy34a

142--Half-inch hole Space Shuttle Endeavour : NASA

--Solar wind map CCMC/Goddard

Lessons Learned from Antarctica

Basic Rocketry

-- Aerojet engine :NASA http://tinyurl.com/pjhaab2

--DeltaV diagram

--Delta V solar system subway map.

A Simple Mathematical Model

-- Author

Chemical Rocket Engines

--Solid fuel pattern diagram http://tinyurl.com/obxf3sd

-- Diagram of J2 engine. NASA. http://tinyurl.com/qhg62fc

--Shuttle Launch. NASA

Slow-speed travel

--Interplanetary Transport Network.. NASA.

--Genesis mission trajectory Wikipedia:NASA/JPL

Nuclear Rocketry

-- Engine. http://tinyurl.com/pnw98wt

--Image of NASA Copernicus (NASA) http://tinyurl.com/bmg8lsd

--Gas core figure http://tinyurl.com/q8thbkf

--Plasma Core Engine. http://tinyurl.com/oyehctt

--JIMO design: Image credit:NASA/JPL

--Copernicus mars transport. Wikipedia: NASA/Pat Rawlings (SAIC))

--Copernicus artwork. NASA

--Gas core nuclear rocket. http://tinyurl.com/q8thbkf

--Nuclear lightbulb engine diagram.

--NASA diagram from PPT

--NASA diagram from PPT

--CERMET engine diagram

--SAFE engine test diagram

-- Improved Atomic Rockets http://tinyurl.com/d3ygk2w

Ion Engines

--Basic diagram

--Deep Space 1 ion engine photo

-- NEXT engine test. Credit: NASA

-- SP-100 reactor http://up-ship.com/blog/?p=23813

--Velocity versus ISP diagram. http://tinyurl.com/qjms54m

--VASMIR engine lab. Sakai/JSC/NASA

-- Artist rendering of the VASIMR powered spacecraft. Ad Astra.

-- Four Busek Hall Effect thrusters. http://www.space.com/21199-space-electric-propulsion-engines.html

--Hayabusha spacecraft. http://tinyurl.com/pbd39q7

Fusion-based engines

--Fusion rocket. George Miley. http://tinyurl.com/oaj8ejw

-- Fusion rocket. U. Washington, MSNW http://tinyurl.com/p6nc3ru

Other Rocket Designs.

---Ikaros solar sail test. Credit:JAXA.

--Lightsail design. The Planetary Society. http://sail.planetary.org/

Bibliography and References

The History of Interplanetary Travel
NASA wants gas stations in space
> http://tinyurl.com/nu3p4zn

Congress wants on-orbit refueling.
> http://tinyurl.com/nowlqr5

Where should we go?
Number of Known Accessible Near-Earth Asteroids Doubles Since 2010
> - http://neo.jpl.nasa.gov/news/news189.html

Why should we go?
Americans want space exploration but not to pay for it -
> http://tinyurl.com/lvvc34x

Tape Recording of meeting between President JohnF. Kennedy and
> NASA Administrator Webb[20]. http://tinyurl.com/plb82rx

Public opinion polls and perceptions of US human spaceflight
> Roger D. Launius, Space Policy 19 (2003) 163–175

Most Americans give low priority to climate change -
> http://tinyurl.com/m7e7u28

A woman sewing a spacesuit.
> http://tinyurl.com/lv9k7t3

History of Spacesuits – NASA –
> http://history.nasa.gov/spacesuits.pdf

How much will it cost?
The space program and US competitiveness (Council on Foreign
> Relations) http://tinyurl.com/nbyjojt

Public Attitudes about space exploration (NSF, 4/2002)
> http://tinyurl.com/nz6mafw

Space settlement a national goal says Congress. (Space News)
> http://tinyurl.com/nmrfm4d

Cruz advocates for space exploration, USA Today (2/24/2015) -
> http://tinyurl.com/nzjt5gg

Interplanetary Travel

Citizens for Space Exploration –
      http://tinyurl.com/pypshcz

The benefits of space exploration (Universe Today) –
      http://tinyurl.com/6kcyhmr

The future of the Space Program – Slate.com (May 2015) -
      http://tinyurl.com/c8j9w8u

China and the next space superpower (Spectrum IEEE)
      http://tinyurl.com/p87t8pe

Americas space future – (Marshall.org)
      http://tinyurl.com/njy8a9o

NASA deep space missions (Orlando Sentinal 4/27/2014)
      http://tinyurl.com/ktulodp

Congress may support renewed space program (Online Athens,
      9/20/20145) http://tinyurl.com/qhenzfd

Votes on 2014 NASA Reauthorization Bill HR 4421.
      http://clerk.house.gov/evs/2014/roll272.xml

NASA defends asteroid mission – Science Times (4/17/2015) -
      http://tinyurl.com/njy4lus

NASA announces degtails on ARM – (Extreme Tech)
      http://tinyurl.com/nsagtos

The Economic Viability of Mars Colonization, Robert Zubrin
      http://www.aleph.se/Trans/Tech/Space/mars.html

Elon Musk and Mars Colonization (Business Insider)
      http://tinyurl.com/pz6ycvy

Colonizing Mars (Red Orbit.com)
      http://tinyurl.com/p4orrg3

Pew Research on Vaccinations and Choice (2/2/2015)
      http://tinyurl.com/psx79wh

The Affordable Care Act and popularity (NPR, 7/12/2012)
      http://tinyurl.com/osso8sz

Gas Tax Hike Widely Unpopular. (The American Consumer, 4/2012)
      http://tinyurl.com/ohq28uj

Pothole damage costs us billions. (WUSA9.com, 2/24/2014)
      http://tinyurl.com/oqzl4yx

US Energy Administration FAQs –
      http://tinyurl.com/k582ttr

Pew Research polling history events –
      http://tinyurl.com/khr6e6s

Interstate Highway History –

http://tinyurl.com/o5hxxlr

Fractured public opinion on infrastructure investment - http://tinyurl.com/pa5nv2u

People don't really know size of NASA budget (The Space Review) - http://www.thespacereview.com/article/1000/1

NASA invests in hundreds of small businesses (NASA, 4/2014) http://tinyurl.com/ooh3l32

US fighter plane 10 years behind (The Daily Beast, 12/26/2014) http://tinyurl.com/mybxudp

F35 fighter too big to kill (Bloomberg.com, 2/22/2013) http://tinyurl.com/ouvowls

NASA 2015 Reauthorization passes House (NASA Watch) http://tinyurl.com/q2lqmdg

The real reason we explore space (Smithsonian Air and Space) http://tinyurl.com/nmprsr6

Science spending in 2015 budget (Science Magazine, 2/2015) http://tinyurl.com/oxvn2gm

How much does NASA cost (US Economy, 2015) http://tinyurl.com/82m4xsg

How shall we explore space? (Universe Today) http://tinyurl.com/on9ymk3

NASA, Mars and dealing with Congress (Vox, 2/4/2015) http://tinyurl.com/nzhvlhn

Some Common Ideas
Stephen Hawkings urges space colonization (Space.com) http://tinyurl.com/qzfhuf3

China to mine moon for helium-3. (Extreme tech) http://tinyurl.com/p8o5n72

Religion and space exploration survey (Space.com) http://tinyurl.com/q7hnw6n

Household debt (Wikipedia) http://en.wikipedia.org/wiki/Household_debt

Components of GDP (About US Economy) http://tinyurl.com/3ejhbh5

Space mining bill passes House (Washington Post, 5/22/2015) http://tinyurl.com/q336krp

How to donate to NASA –

http://spaceindustrynews.com/how-to-donate-to-nasa/
E-commerce sales forecast (Statista.com)
    http://tinyurl.com/q3cjotd
Five things people want to buy on the internet (Insitesoft/.com)
    http://tinyurl.com/nbvfdt4
Household debt in the USA (Washington Post, 8/15/2013)
    http://tinyurl.com/m9xossp
ISS module relocated to make way for Bigelow module (NASA)
    http://tinyurl.com/losjsrr

Resources and Mining
The composition of the asteroid belt – (Astrobites, 2/1/2014)
    http://tinyurl.com/pjru78q
The composition of meteorites.
    http://www.permanent.com/meteorite-compositions.html
Gold in meteorites and Earth's crust (USGS Report)
    http://pubs.usgs.gov/circ/1968/0603/report.pdf
Composition of iron meteorites – Andrew Campbell (U. Chicago)
    http://tinyurl.com/o5pqeam
NASA's NEAR mission (Science.NASA)
    http://science.nasa.gov/missions/near/
Eros Fact Sheet (NASA)
    http://nssdc.gsfc.nasa.gov/planetary/text/eros.txt
NASA's asteroid capture mission. (The Times, May, 2013)
    http://tinyurl.com/bpy73eq

The Lunar Surface
Chemical Composition of Lunar Meteorites and the Lunar Crust
    http://www.kurat.at/pdf/492.pdf
Gillis, J. J., Jolliff, B. L., and Korotev, R. L. (2004) Lunar surface
    geochemistry: Global concentrations of Th, K, and FeO as
    derived from Lunar Prospector and Clementine data.
    Geochimica et Cosmochimica Acta, v. 68, p. 3791-3805.
Planetary Crusts: Their composition and origin – Stuart Taylor and Scott
    McLennan (Cambridge Planetary Science)
    http://tinyurl.com/ohkmtxu
NASA looking for commercial mining companies (The Verge, 2/9/2014)
    http://tinyurl.com/p4qlu8q

Bibliography and References

The Lunar Environment – David Beale ESMD Course.
  http://tinyurl.com/qarrdrs
Caterpillar and NASA team for mining (IB Times)
  http://tinyurl.com/k7k8lbt
Radar finds ice on moon (NASA)
  http://tinyurl.com/yfx7xdp
Could Helium-3 solve Earth's energy problems? (IO9, 5/11/2012)
  http://tinyurl.com/oejosuj

Mercury Resources

Gravity Field and Internal Structure of Mercury from MESSENGER,
  10.1126/science.1218809
Mercury surface resembles rare meteorites – (Space.com)
  http://tinyurl.com/qzu4sf5
MESSENGER FAQs (NASA/JHUAPL)
  http://messenger.jhuapl.edu/qa/?faq=1&ca=38#qn605
Mercury surface composition from MESSENGER data
  http://onlinelibrary.wiley.com/doi/10.1029/2012JE004153/full

Mars Resources

Mars minerals of commercial importance
  http://tinyurl.com/p66s58w
Nitrates discovered in Martian soil by Curiosity (Spaceflight, 4/8/2015)
  http://tinyurl.com/pxr7cj4

Asteroids, Comets and Moons

The composition of Comet C/2009P1 Table
  http://tinyurl.com/nfhp9rq
Surface composition of Comet Wild 2
  http://stardust.jpl.nasa.gov/news/news110.html
McCord, T. B. and others, 1998, Salts on Europa's surface. Science, v.
  280, p. 1242-1245.
Surface composition of Iapetus
  http://tinyurl.com/qgosc9b
Titan atmosphere composition
  http://fas.org/irp/imint/docs/rst/Sect19/Sect19_19.html
Trace gases in Earth's atmosphere
  http://tinyurl.com/palz9te

Interplanetary Travel

Vesta minerology and composition
> http://tinyurl.com/p6pbfby

Composition of asteroids
> http://tinyurl.com/oaer874

Composition of Comet 67P
> http://www.hou.usra.edu/meetings/lpsc2015/pdf/2092.pdf

Tourism

Space-X designing a competitor to NASA's SLS. (NASA, 8/29/2014)
> http://tinyurl.com/pgwjcgz

Space-X Mars Sample-Return Mission (Space.com, 3/7/2014)
> http://tinyurl.com/qd7l8pq

When should we go?

NASA announces plans for Mars landing (Salon, 4/9/2014)
> http://tinyurl.com/qdqn87n

How NASA plans to get to Mars. (Planetary Society, 11/19/2014)
> http://tinyurl.com/q7dj73r

China Space Station Plans, (Space.com)
> http://www.space.com/27440-china-space-station-plans.html

Bold new missions to Jupiter and Saturn planned (Space.com, 2/18/2009)
> http://tinyurl.com/qcg57df

Missions proposed to explore Uranus (Space.com, 10/13/2011)
> http://tinyurl.com/5vvx5js

NASA's golden age coming to an end. (Vox.com, 2/23/2015)
> http://tinyurl.com/pxjvxoj

Europa mission gets White House approval, (Space.com, 2/2/2015)
> http://tinyurl.com/lcjbd55

NASA mission to explore Titan (Clapway.com, 4/4/2015)
> http://tinyurl.com/q3jnnn9

NASA designs sub for Titan seas. (Discovery, 2/12/2015)
> http://tinyurl.com/mkgbcod

NASA's FY16 budget
> http://tinyurl.com/neh3tez

NASA planetary science report (NASA, 4/30/2015)
> http://tinyurl.com/npjvdmp

House Committee Passes NASA Authorization, (Geological Society)
> http://tinyurl.com/nkkymkk

Bibliography and References

NASA 2016 budget, (The Planetary Society)
> http://tinyurl.com/nnfwmrg

JPL studies low cost Mars missions (Space News)
> http://tinyurl.com/or8wam3

NASA planning humans to Mars by 2030s (The Standard Daily)
> http://tinyurl.com/ojungkq

Places to Visit

Deep space outpost near moon.
> http://tinyurl.com/7cfvpb7

ESA Europa mission in 2022.
> http://tinyurl.com/pjac49v

Uranus Pathfinder mission
> http://tinyurl.com/oy6hd8p

Interplanetary Mission analysis and Design 2006
> http://tinyurl.com/pdsbyc4

Uranus radiation belts.
> http://tinyurl.com/oa43p9j

Triton coldest spot in solar system (NY Times, 8/29/1989)
> http://tinyurl.com/om3kvsu

Stone, E. C. and Cummings, A. C. and Looper, M. D. and Selesnick, R. S. and Lal, N. and McDonald, F. B. and Trainor, J. H. and Chenette, D. L. (1989) Energetic Charged Particles in the Magnetosphere of Neptune. Science, 246 (4936). pp. 1489-1494.

Orbital Platforms

NASA targets moons of Mars for robotic mission (Space.com, 4/3/2014)
> http://tinyurl.com/khawlbd

Making a case for a mission to Phobos (Universe Today, 10/1/2014)
> http://tinyurl.com/k595uxl

New Phobos sample return missions:
> http://www.csc.caltech.edu/talks/hopkins.pdf

Interplanetary communication.

10 kW laser from IPG Photonics
> http://tinyurl.com/p6kzz43

Galaxy 13 information

Interplanetary Travel

http://tinyurl.com/ogsemqs

International satellite industry price per pound to orbit, (Futron)

http://tinyurl.com/oykba8k

Interplanetary laser communication

http://tinyurl.com/phnrkyd

NASA to test laser communication

http://tinyurl.com/om4dgd2

NASA to test interplanetary internet, (White House)

http://tinyurl.com/qxes6e6

Radiation Hazards

Want to colonize an alien planet? Send 40,000 people. (Space.com)

http://tinyurl.com/m4apt2e

Human voyages to Mars pose higher cancer risks. (Universe Today)

http://tinyurl.com/lbfjzok

Would a trip to Mars damage your brain?, 2011, Jim Schnabel,

http://www.dana.org/News/Details.aspx?id=43154

Mission to Mars could mess with your brain, 2013, Ian O'Neill,

http://tinyurl.com/p2w8gay

How much radiation will the settlers be exposed to?, 2015, MarsOne,

http://tinyurl.com/lf5v6rj   and, http://tinyurl.com/kyfhvfv

Rapid Diagnosis in Populations at Risk from Radiation and Chemicals,

http://tinyurl.com/n75lzjh

Health effects of nuclear radiation, 2015, Jerry Cuttler,

http://tinyurl.com/nh44ebl

Radiation hazard of relativistic interstellar flight, O. Semyonov

http://tinyurl.com/ncqponp

Lessons Learned from Antarctica

Travel to Antarctica rates.

http://tinyurl.com/o8kancl

The Hazards of Interplanetary Space

Basic Rocketry

Rocket propulsion methods (Wikipedia)

http://tinyurl.com/omfob9l

Chemical Rocket Engines

Math problem about rocket properties, Isp, mass flux, speed.
    http://spacemath.gsfc.nasa.gov/weekly/10Page76.pdf

Slow-speed travel

Interplanetary superhighway (JPL)
    http://tinyurl.com/oflngde

Nuclear Rocketry

Advanced propulsion systems
    http://tinyurl.com/pnw98wt
Nuclear Thermal Propulsion, NASA, Stanlet Borowski
    http://tinyurl.com/pcrl4s9
Nuclear Thermal Rocket characteristics (NASA, Borowski)
    http://tinyurl.com/pflfs7f
Is the future of space nuclear? (The Daily Mail, 2/3/2015)
    http://tinyurl.com/namnfan
The limits of solid core nuclear rockets, (NASA Spaceflight)
    http://tinyurl.com/obndere
Nuclear Rockets (Course lecture, Craig Kluever)
    http://tinyurl.com/no3feb4
Thode, L., Cline, M., Howe, S. (July–August, 1998). Vortex formation and stability in a scaled gas-core nuclear rocket configuration. Journal of Propulsion and Power. Pg. 530-536.
Gas core nuclear rocket design,
    http://tinyurl.com/q7hm2os
Plasma core rocket and MHD vortex
    http://tinyurl.com/oyehctt
Interplanetary trajectories,
    http://tinyurl.com/nwh8s3f
The basics of nuclear rocketry, (LANL)
    http://tinyurl.com/ndbs5dy
Ultra high specific impulse nuclear propulsion, (Dept of Energy)
    http://tinyurl.com/pob4yq3
Future Spacecraft Propulsion Systems (Claudio Bruno, Paul Czysz)
    http://tinyurl.com/q3sfsyp
Nuclear thermal rocket simulation in NPSS (NASA)

http://tinyurl.com/pxyyqho

A technical and economic introduction to nuclear rockets,
http://tinyurl.com/pg7wjp2

Nuclear propulsion choices for space exploration (McLaren and Ragheb)
http://tinyurl.com/ocqymub

Fission power back on NASA's agenda (Nature, 2/6/2012)
http://tinyurl.com/76snk5j

Reviewers caution NASA about nuclear craft, (Nature, 9/9/2004)
http://tinyurl.com/nwkmu8n

Ground testing a nuclear thermal rocket, Howe, (AIAA, 2012)
http://tinyurl.com/p5cm4py

NASA's NTREES facility, (NASA)
http://tinyurl.com/n5mvrnn

Reducing the Risk to Mars: The Gas Core Nuclear Rocket' by Howe, DeVolder, Thode and Zerkle,
https://www.fas.org/sgp/othergov/doe/lanl/la-ur-97-3361.pdf

Ion Engines

An overview of the NEPSTP satellite
http://adsabs.harvard.edu/abs/1994iece.confR...7V

IAA Commission 3 – Nuclear propulsion
https://iaaweb.org/iaa/Studies/nuclearpropulsion.pdf

VASMIR vX-2000 meets milestone.
http://tinyurl.com/njtekz3

Variable Isp trajectories for solar system exploratiuon
http://tinyurl.com/nmaf85b

New types of propulsion in space
http://www.brighthub.com/science/space/articles/71322.aspx

Commercial plasma rocket reaches 200 kW milestone,
http://tinyurl.com/yatbroa

VASIMR rocket will take astronauts to Mars in 39 days,
http://tinyurl.com/pulsckm

The VASIMR Hoax, Robert Zubrin (Space News)
http://tinyurl.com/oqjbr7s

Zubrin claims VASIMR is a hoax (Universe Today, 7/13/2011)
http://tinyurl.com/6layg8b

High specific impulse nuclear-electric missions, (International Astronautical Federation, 2005), http://tinyurl.com/qhhlfff

Bibliography and References

Miniature nuclear reactors for sale! (Daily Tech, 11/10/2008)
  http://tinyurl.com/6lj95n and http://tinyurl.com/qfc4qbx

NASA SBIR Proposal: Advanced Photovoltaic Systems with >1,000 W/kg
  efficiency and < $50/W Cell Cost http://tinyurl.com/qbxo3f2
NASA eyes ion engines for mars orbiter mission in 2022.
  Spaceflight Now, 2015, http://tinyurl.com/o5oc2qp

Fusion-based engines
The fusion-driven rocket (NASA 3/27/2012)
  http://tinyurl.com/bp726a9
Spherical torus nuclear fusion propulsion (NASA/AIAA)
  http://tinyurl.com/qd6ktma
Direct fusion drive for advanced space missions (Princeton)
  http://www.psatellite.com/research/DFDWhitepaper.pdf
The Viper pulsed fusion rocket
  http://tinyurl.com/oaj8ejw
Fusion rocket could take astronauts to Mars in 30 days,
  http://www.technewsworld.com/story/77758.html
Fusion based nuclear propulsion seminar report,
  http://tinyurl.com/ogu7jeg
The case and development path for fusion rockets,
  http://tinyurl.com/nc7rps4
Nuclear fusion rockets may be key to exploring deep space, (NBC
  news,6/11/2013) http://tinyurl.com/p6nc3ru

www.ingramcontent.com/pod-product-compliance
Lightning Source LLC
Chambersburg PA
CBHW051849170526
45168CB00001B/39